DEVELOPMENTS IN PVC PRODUCTION AND PROCESSING—1

DEVELOPMENTS IN PVC PRODUCTION AND PROCESSING—1

Edited by

A. WHELAN and J. L. CRAFT

National College of Rubber Technology, Holloway, London, UK

APPLIED SCIENCE PUBLISHERS LTD
LONDON

APPLIED SCIENCE PUBLISHERS LTD
RIPPLE ROAD, BARKING, ESSEX, ENGLAND

ISBN: 0 85334 741 7

WITH 26 TABLES AND 76 ILLUSTRATIONS

Printed in Great Britain by Galliard (Printers) Ltd, Great Yarmouth

PREFACE

Polyvinylchloride is one of the major thermoplastics materials because it is a polymer which is relatively cheap and versatile. It has been estimated that in 1976 over eight million tons of PVC were used throughout the world in a bewildering number of products. PVC has therefore developed in the last thirty years from being a dubious substitute for natural materials to a high tonnage commodity in its own right. Its usage continues to grow despite handicaps such as thermal instability and residual vinyl chloride monomer. These and other subjects will be covered in the chapters which follow.

This book is not intended as a complete guide to PVC, but rather its purpose is to review some of the more recent developments that have occurred in the world of PVC. It is hoped that this book will therefore be of particular interest to those who are concerned with the preparation, processing and application of this, the most versatile thermoplastics material that we know.

All of the contributors are colleagues from industry and are specialists in this particular field. We would like to express our thanks to them and their companies for their participation in the preparation of this book. Grateful thanks are also tendered to the Governors of the Polytechnic of North London for allowing us to act as editors and for providing the facilities without which the book could not have been produced.

CONTENTS

LIST OF CONTRIBUTORS

H. ADAM

Werner & Pfleiderer (UK) Ltd, 72 Compstall Road, Romiley, Stockport SK6 4DE, UK.

G. P. BARNETT

Product Manager, Extrusion Machinery Division, Francis Shaw & Company Ltd, PO Box 12, Corbett Street, Manchester M11 4BB, UK.

R. H. BURGESS

Imperial Chemical Industries Ltd, Plastics Division, PO Box 6, Bessemer Road, Welwyn Garden City, Herts AL7 1HD, UK.

H. M. CLAYTON

Technical Manager, PVC Division, British Industrial Plastics Ltd, Aycliffe Industrial Estate, Darlington, Co. Durham DL5 6AN, UK.

K. T. COLLINGTON

Agrochemical Division, Fisons Ltd, Hauxton, Cambridge CB2 5HU, UK.

R. A. ELDEN

Sheet & Film Division, British Industrial Plastics Ltd, Brantham, Manningtree, Essex CO11 1NJ, UK.

J. PICKERING

Technical Director, Fibrenyle Ltd, Skylon House, Ellough Industrial Estate, Beccles, Suffolk NR34 7TB, UK.

J. B. PRESS

Technical Director, Wavin Plastics Ltd, Brunswick Road, Cobbs Wood, Ashford, Kent, UK.

P. RICE

Werner & Pfleiderer (UK) Ltd, 72 Compstall Road, Romiley, Stockport SK6 4DE, UK.

D. A. TESTER

Imperial Chemical Industries Ltd, Plastics Division, PO Box 6, Bessemer Road, Welwyn Garden City, Herts AL7 1HD, UK.

W. V. TITOW

Yarsley Research Laboratories Ltd, The Street, Ashtead, Surrey K21 2AB, UK.

D. A. TREBUCQ

Wavin Plastics Ltd, PO Box 12, Hayes, Middlesex UB3 1EY, UK.

Chapter 1

THE PATTERN OF USAGE OF PVC

D. A. TESTER

ICI Plastics Division, Welwyn Garden City, UK

SUMMARY

This review sets out to examine the current level and pattern of PVC consumption in the UK, and to consider future growth, in terms both of general economic factors and particular application. Whilst technical innovations are discussed and considered to be of importance, the predicted future of PVC is seen to depend far more on the growth or stagnation of the general economy than on any particular development.

The commercial competitiveness of PVC against other plastics and traditional materials is assessed in the light of the price rises through this decade. In general, it is concluded that the price indices of all those materials have increased similarly and that the competitiveness of PVC has not been significantly weakened during this period of exceptional inflation. The impact of environmental concerns on the consumption of PVC and on future trends is also discussed.

1.1 INTRODUCTION

A review of the growth and pattern of polyvinylchloride (PVC) consumption in the United Kingdom becomes, in effect, a review of the whole economy. The reasons for this are not hard to find. Firstly, PVC has grown in the last thirty years from a dubious substitute for natural materials to a high tonnage commodity in its own right. Secondly, it finds application in building, transport, communications, packaging—in fact in every major aspect of our economic activity. This, in turn, springs from the inherent versatility of the polymer, giving rigid or flexible compositions, transparent or opaque, capable of being fashioned into such diverse products as cables, pipes, floor tiles, bottles, ranch fencing and surgical gloves. To the reviewer this presents the advantage of a broad-based technology for discussion. However, there is the corollary that very few innovations, whatever their technical merits, can now be expected to command a tonnage sufficient to change decisively the growth rate or pattern of usage. In practice, the recent history and predicted future of PVC depend far more on the growth or stagnation of the general economy than on any particular development, as will be evident from the following pages.

1.2 PREDICTIONS FOR THE FUTURE

In a previous publication,[1] it was stated that PVC consumption in the UK, having grown from around 1000 tonnes per annum in the early 1940s was expected to reach 700 000 tonnes in 1980. It was considered by the author that 'this growth prediction is realistic and some dramatic recession in our economy would be required to cause it to be very wrong'. That was written early in 1973, and as we all know, a dramatic recession in our economy did subsequently occur. 1974, with industrial problems and, for a while, raw material shortages, saw little growth in PVC consumption, and following the impact of quadrupled oil prices, 1975 saw a marked recession in the western economies, and therefore in PVC usage. There are reasonable hopes that the economic recovery begun in 1976 will continue, particularly as North Sea oil begins to make its impact, and that a steady increase in PVC consumption will resume, but the lost ground of the mid-seventies will not be made up, at least in the short term. UK predictions for 1980 are now around 550 000 tonnes, as shown in Fig. 1, compared with the 700 000 tonnes previously expected.

The earliest applications of PVC were in its plasticised flexible form, but with progressive improvements in technology, the processing of unplasticised PVC was increasingly exploited, bringing faster growth in rigid applications. It was predicted in 1973 that by 1976 rigid applications would match flexibles in tonnage in the UK, and become increasingly dominant in subsequent years. In the event, the 1974–76 period saw a 'freezing' of the relative consumptions in rigid and flexible applications. This is largely because the rigid applications, particularly pipe and conduit, depend so heavily on the building construction industries, which have been hard hit by recession and public expenditure curbs. It now seems likely that

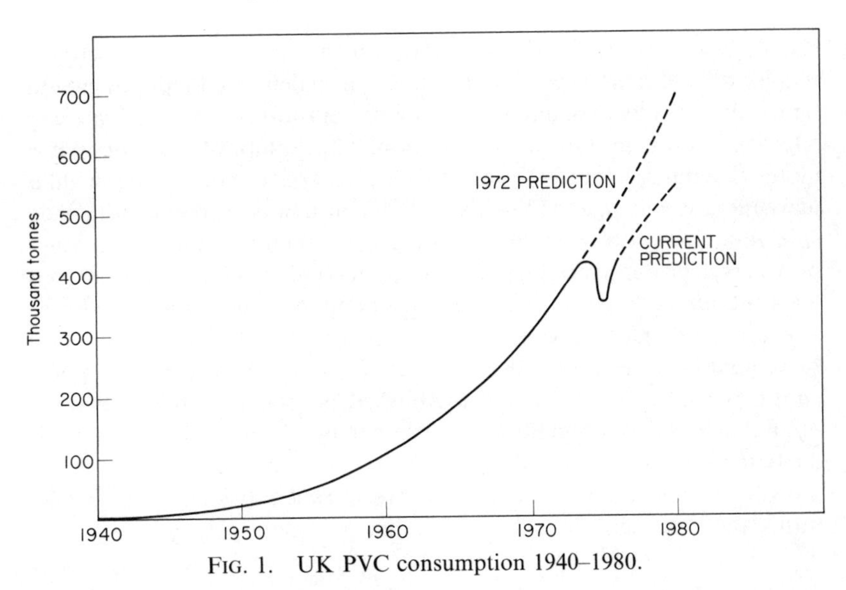

FIG. 1. UK PVC consumption 1940–1980.

it will be 1980 before rigid applications consume more PVC than flexibles, but the faster growth in rigid products should be increasingly apparent thereafter.

1.3 THE RELATIVE IMPORTANCE OF PVC

Table 1 shows the UK consumption of large tonnage plastics (excluding the rigid thermosetting resins) for the last two years. It will be seen that PVC had the highest consumption on a tonnage basis, as it has throughout this

TABLE 1
UK CONSUMPTION OF PLASTICS (THOUSAND TONNES)

Product	1975	1976
PVC	345	422
LD polythene	318	417
HD polythene	90	120
Polypropylene	156	185
Polystyrene	123	160
ABS	26	36
Polyurethanes	60	71

decade, with low density (LD) polyethylene a close second (and actually first in volume terms because of the difference in density). Predicted growth up to 1980 is likely to maintain these relative positions, with both reaching 500–550 000 tonnes annual consumption. The pattern of use is of course quite different, with 75% of LD polyethylene consumed in film manufacture, whilst, as will be seen, PVC consumption covers a very wide spectrum, but with over 20% in pipe and conduit. Among the other polymers, high density (HD) polyethylene and polypropylene have enjoyed the most rapid growth in recent years, doubling consumption since 1970. As a relative newcomer, polypropylene is expected to expand in consumption by around 14% per annum over the next few years, against the 7–8% now considered most likely for well established polymers. Although there is undoubtedly some competition between polypropylene and other plastics, including PVC, it is expected that the bulk of its expansion will be through growth and innovation, e.g. in films, fibres and large structural foam items, rather than through replacement of other polymers.

1.4 THE PATTERN OF CONSUMPTION

A comprehensive list of PVC applications would be very long indeed, but the abbreviated list in Table 2 includes the major tonnage applications, and those that have shown fastest growth in recent years. The table indicates the growth of each area from 1967. Complete statistics for 1976 were not available at the time of writing, but all the indications are that the levels and pattern of consumption were close to those of 1974, rather than to the atypical performance of 1975.

TABLE 2
UK CONSUMPTION OF PVC (THOUSAND TONNES) BY APPLICATION

Product	1967	1972	1974	1975
Flexible film and sheet	38	45	47	33
Rigid foil and sheet	8	20	30	21
Bottles	1	10	14	12
Records	9	16	21	17
Rigid extrusion	35	90	104	90
Cables	36	42	50	46
Flooring	31	28	30	26
Fabric and paper coating	17	34	35	28
Footwear	8	15	18	15

Taking the applications in order, flexible film and sheet represents an area of considerable, but slow growing, tonnage. The minor portion taken up by thin extruded packaging film is growing quite fast, but the market for thicker calendered sheet, one of the longest established of PVC products, can only be expected, at best, to match the growth rate of the gross national product. In contrast, the consumption of rigid foil and sheet has been increasing rapidly in recent years and this growth pattern is expected to continue. This category really includes two separate products, the relatively thin foil used in thermoformed packaging, and the thicker sheet, flat or corrugated, employed in roofing and glazing applications.

Bottles represents a comparatively new application for PVC, which grew in spectacular fashion in 1967–72, but is, inevitably, now expanding at a slower pace. The manufacture of gramophone records consumes an appreciable tonnage of PVC, growing at 10–20% per annum up to 1975. Continuing growth is expected, unless or until the popularity of cassettes creates severe inroads into potential sales of records.

Rigid extrusion will be seen to have grown to the point of dominating the pattern of consumption, with a usage of around 100 000 tonnes per annum. Although this area includes a wide range of extruded products, the major proportion by far is made up of pipe and conduit. Whilst the market for rainwater goods has been completely taken over by PVC, and high penetration of the pressure pipe market has been achieved, penetration of the very large market in domestic drainage is as yet quite modest. Here, the future rate of penetration will depend very critically on the effective price in each pipe size, compared with that of glazed clay, the traditional material.

The cable market is the second largest application area for PVC, but one in which penetration of the most technically suitable applications, e.g. housewiring, has long been completed, and future growth must closely follow that of building construction, and to a lesser extent, the car industry.

A few years ago, the PVC flooring industry appeared to be in decline. Continuous sheet flooring and the ubiquitous PVC tile had both enjoyed rapid growth in the 1960s, but were giving way to more luxurious flooring, particularly low priced tufted carpets. However, in the last few years, the advent of 'cushionfloor' (see following section) has revitalised the flooring market, though partly at the expense of traditional PVC flooring. The result is likely to be a fairly static consumption for flooring as a whole, in future years, but with steady growth for the cushionfloor segment of the market.

Fabric and paper coating covers two very different applications, though with similar production technology. Fabric coating, to produce 'leathercloth' products, is a long established PVC application, exploited in domestic upholstery, car seats, handbags, and travel goods. As such, it is vulnerable to competition from other materials, such as urethanes and synthetic woven fabrics, and also highly sensitive to the general level of the economy. Paper coating refers mainly to the production of vinyl wall coverings, which have been growing rapidly in popularity, partly at the expense of traditional wallpaper. In this area, a high growth rate is predicted for some years to come.

Finally, footwear denotes an area of flexible moulding where there is now competition from polyurethane and thermoplastic rubbers, but where the real threat to future consumption lies in the possible decay of the whole UK footwear industry, in the face of relentless imports.

1.5 GROWTH THROUGH INNOVATION

Since over a quarter of the total UK consumption of PVC is in rigid extrusion, the future must depend critically on growth in this area. The dominant tonnage, as already stated, is in pipe and conduit. Here, there is considerable growth potential for PVC in domestic drainage, where clay pipe still serves by far the major part of the market. The price of clay pipe has risen less than that of many other materials in the last two years, and as a result the basic price is now somewhat lower than that of PVC pipe, even in smaller diameters. Whilst there are installation cost advantages for PVC, the rate of penetration into this very large market in the UK and elsewhere must depend critically on price movements in the coming years. This has given impetus to the investigation of methods for the production of cheaper PVC drainage pipe, e.g. by the use of ribbed structures with lower weight per unit length for a given stiffness. Two companies (Kamphuis in Holland, and Lundberg in Sweden) have developed systems for extruding ribbed pipe, whilst Durapenta in South Africa has developed an injection moulding technique for producing continuous lengths of ribbed pipe. At the time of writing, these systems have not been exploited in this country. It is too early yet to gauge the commercial importance of such techniques, but it may be expected that they would be a lot slower than conventional extrusion, and must involve investment in new fabrication plant.

Rigid profiles and sections find a variety of applications in building and furnishing, where there is considerable potential. Penetration of the UK window frame market has been slow compared with that in some European countries, but several companies are now importing European PVC window frame systems, with a view to future manufacture in the UK. In

FIG. 2. The 'Grippa-Frame' system.

recent years PVC has been effectively exploited in double glazing systems for the improvement of existing homes. Figure 2 shows the 'Grippa-Frame' snap-fitting PVC system, extruded by Markim Extrusion for Oxford Double Glazing Ltd. In furnishing, rigid PVC foil is used for edge trim, particularly on chip board, and in recent years rigid PVC has been increasingly employed for the extrusion of drawer sections. The 'Sheer-Glide' system, manufactured by LB Plastics Ltd, combines rigidity and economy of material, by the use of hollow sections, and is widely used in fitted furniture for kitchens and bedrooms.

Some years ago, it was considered that rigid PVC foams in extruded form would find considerable application in building and furnishing applications, as a substitute for soft wood. These foams are certainly wood-like in density and workability but with greater dimensional stability, and have, of course, the ability to be extruded into the desired profiles with minimal waste and no surface finishing requirement. By 1972, 20 000 tonnes of PVC were consumed annually as rigid foam in the USA, and development projects were under way in this country. Since then increases in the price of PVC, relative to timber and the general recession in building, have slowed the growth of rigid foam applications. Nevertheless, several companies in the UK remain actively interested, and with the increasing costs of the fabricating and finishing processes inherent in the use of wood, it may be predicted that rigid PVC foams will play a significant role in the future.

Rigid PVC has been traditionally considered to be an exacting material for the injection moulder, compared with, say, polystyrene. Apart from the moulded fittings essential to pipe manufacture, there have been only very modest tonnages of PVC consumed in rigid injection moulding. However, with improvements in formulation technology and the increasing expertise of the injection moulder, it is now possible to exploit the product properties of rigid PVC in a wide range of mouldings. A current example is the PVC chassis runner incorporated in the GEC colour TV set. Ten of these runners, illustrated in Fig. 3, act as struts, spacers, and slotted supports for printed circuit boards. In contrast to this relatively small but intricate moulding is the robust Molyneux drainage gully, shown in Fig. 4.

In the previous section it was mentioned that the consumption of PVC in flooring had begun to decline in recent years, but that the advent of 'cushionfloor' was providing a popular innovation in this area. Cushion flooring products are now promoted by all the major manufacturers of PVC flooring, and are probably familiar to most readers. The essential feature is the inclusion of a foamed 'cushion' between the base of the flooring and the tough transparent wear layer. The foam layer carries

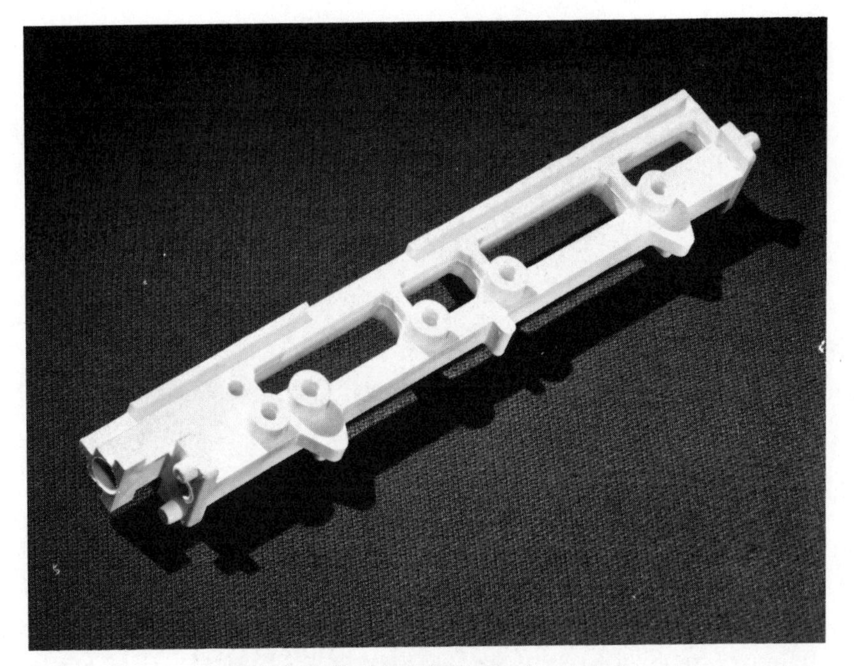

FIG. 3. Moulded PVC chassis runner.

a printed pattern and the wear layer is generally embossed in register with this. Continuous PVC flooring has thus acquired something of the resilience and feel of luxury associated with tufted carpets and carpet tiles, whilst retaining the hard wearing, easy cleaning characteristics of the solid PVC floor. This aplication, like the vinyl wall coverings, which are still growing in popularity, employs a 'paste' polymer, a grade of PVC which, with plasticiser, forms a smooth plastisol for spreading and subsequent gelation. These two applications should assure above average growth potential, for this type of polymer, into areas that are very much in the public eye.

Motor car manufacture provides a sensitive barometer for the fluctuations of our economy, and is a very large consumer of a wide range of materials. The applications for PVC are generally unspectacular, but varied and essential. They range from car harness wiring to trim, sealant, roof lining and, of course, seat upholstery—though, as already mentioned, this leathercloth application faces strong competition from brushed nylon fabrics. A new and attractive development is the use of injection moulded PVC, over a steel core, for the manufacture of steering wheels. These have

FIG. 4. Moulded PVC drainage gully.

the feel and grip of soft leather, with tough hard wearing properties. Figure 5 shows the PVC steering wheel chosen for the Vauxhall Chevette.

Over the last few years there has been considerable concern over the toxicity of vinyl chloride, the monomer precursor of PVC, and its implications for the use of PVC in packaging. The nature of this problem and the response of the industry are discussed in detail in Chapter 3 of this book. Here we need only remark that the use of PVC in packaging in the UK and Western Europe has not only survived but seems likely to sustain a steady growth in future. The uses of PVC film, foil and bottles are too well known now to require further comment, but it is refreshing to note the increasing exploitation of designs that are relatively easy to achieve in blow-moulded plastic but more difficult and expensive in traditional materials.

FIG. 5. PVC steering wheel, injection moulded over a steel core.

This is exemplified by the two-litre beer jug introduced by Greene King, shown in Fig. 6.

1.6 UK CONSUMPTION IN A WORLD CONTEXT

It will be apparent that the consumption of PVC in this country has reached a very considerable tonnage, across a wide field of application, but the consumption per head is, nevertheless, modest by comparison with some other communities. In fact, we come bottom of the 'league table' among major industrial countries (excepting the USSR) in times both of growth and of recession, as shown in Table 3.

FIG. 6. Two-litre blow-moulded PVC beer jug.

TABLE 3
ESTIMATED PVC CONSUMPTION (KG/HEAD)

	1971	1974	1975
West Germany	13·4	15·2	12·7
France	8·9	12·1	10·4
Japan	8·8	11·4	9·5
USA	6·8	9·5	7·6
Italy	6·3	8·6	6·4
UK	5·9	7·3	6·3

The reasons are complex, and depend not only on special circumstances, such as the enormous consumption of bottled water in PVC in France and the scarcity of homegrown timber in Japan, but also on the innate conservatism of some British industries, particularly in building. Whilst there is no likelihood of a dramatic change in relative national consumptions in the near future, the figures do illustrate the potential for growth of PVC usage in the UK, without presupposing any extraordinary changes in life style or prosperity.

1.7 OTHER FACTORS INFLUENCING GROWTH

1.7.1 Economic Considerations

Accepting that the economic fortunes of PVC must broadly follow those of the general economy, and that prices must inevitably rise with continuing inflation, will the price of PVC increase faster than that of other, competitive, materials? If we consider first the large tonnage plastics, it is instructive to examine the price movements for general purpose grades from 1970 to 1976, shown in Table 4.

It will be seen that prices were generally stable or, indeed, falling at the beginning of this decade, but that they increased markedly in 1974–76. Broadly speaking, however, the relative position in 1976 was not very different from that of 1972. The exception is polypropylene, where the increase has been proportionately rather less. In 1970–72, polypropylene was at an early stage of its growth curve as a high tonnage plastic, and could still be expected to benefit from economics of increasing scale, offsetting to some extent the increasing costs of monomers and fuel.

It has been pointed out in an earlier section that clay pipe has increased in price less in recent years than equivalent PVC pipe, with implications for the

TABLE 4
AVERAGE PRICE INDICES OF PLASTICS POLYMERS TO LARGE CUSTOMERS (1970 = 100)

Year	*PVC*	*Low density polythene*	*High density polythene*	*Polypropylene*	*Polystyrene (GP)*	*ABS*
1970	100	100	100	100	100	100
1971	100	100	91	100	100	100
1972	100	100	83	100	96	100
1973	109	108	89	100	100	104
1974	164	204	149	136	223	202
1975	214	208	171	141	235	224
1976[a]	264	262	197	170	254	260

[a] Average for first half of year.

penetration of PVC into the drainage market. It would be quite untrue, however, to assume that this was part of a general trend in competitive prices. Although polymer prices are highly sensitive to increases in the cost of oil-based feedstocks, many other materials are highly sensitive to concurrent rises in fuel and transport costs, and make greater demands on highly paid labour in their processing and fabrication.

Table 5 shows the movement in wholesale price indices 1970–76, for a range of materials and products relevant to PVC consumption. Whilst it is unwise to draw too many conclusions from a list that includes both raw materials and semi-finished items, it is nevertheless evident that, with the single exception of copper pipe, all these materials, like PVC, have increased in price by around 150–200% in this period.

Whilst no plastics can be considered today to be exceptionally cheap raw materials, as they may have been in the 1960s, the unprecedented increase in oil prices has not, in general, weakened the competitiveness either of plastics as a whole, or of PVC in particular. For the future, it is expected that PVC prices will increase very much in line with the general rise in the cost of living.

1.8 ENVIRONMENTAL CONCERNS

1.8.1 Monomer Toxicity

During the last few years there has been considerable concern over the toxicity of vinyl chloride and its implications for PVC production, fabrication, and application in sensitive areas, particularly food packaging.

TABLE 5

WHOLESALE PRICE INDICES OF SELECTED MATERIALS AND PRODUCTS (1970 = 100)

Year (average)	*Cement*	*Precast concrete pipe*	*Iron pipe and fittings*	*Steel tube and fittings*	*Copper tube*	*Imported soft wood*	*Paper and board*	*Glass containers*	*PVC*
1970	100	100	100	100	100	100	100	100	100
1971	120	115	109	111	86	104	106	115	100
1972	127	119	118	120	86	112	111	119	100
1973	127	125	127	128	120	196	120	122	109
1974	146	146	170	179	137	259	171	150	166
1975	204	212	239	241	109	238	219	204	213
1976[a]	250	266	286	284	153	315	240	234	273

[a] As at June 1976.

PVC, like most high polymers, is itself non-toxic, but in its production small quantities of unreacted monomer remain in the polymer, and minute traces can persist in a fabricated PVC product. Vinyl chloride (VCM) had been traditionally accepted as presenting no health hazard, at least in the trace amounts remaining in fabricated PVC. However, in January 1974, a link between the inhalation of VCM and angiosarcoma in man was confirmed, and the UK PVC producers embarked on programmes to drastically reduce the levels of residual monomer in the working environment, the polymers, and the fabricated products, particularly food packages. The results represent a remarkable technical achievement, though at some considerable cost, by the whole industry. This topic is discussed in detail in Chapter 3, and is only considered here in terms of its effects on PVC consumption. The most serious consequence would, of course, have been the curtailment of PVC production, with resultant shortages and an enforced shift to other materials. The dominant and even unique role of PVC in some applications, and the importance of the industries involved, in terms of both jobs and investment, pointed to the need for a constructive approach. In the event, progressive reduction in atmospheric monomer levels in PVC production plants has been achieved in the UK without serious reductions in capacity. At the outset, there seemed to be a real possibility of curtailment of PVC usage in the packaging of food and drink, since through this application VCM could be ingested by the general public. However, over the last two years, progressive reductions in residual monomer in packaging polymers and feedstocks have made it possible to set a limit of 1 part per million in PVC bottles, where contact with liquid foods had given rise to most concern over extraction. At this level in the bottle, the levels in the contents after prolonged storage are extremely low—invariably less than 50 parts per billion (10^9), and typically less than 10 ppb.

Undoubtedly, certain potential developments in PVC packaging were shelved, or at least postponed, because of VCM concerns, but main stream packaging applications in the UK have not suffered significantly, and as mentioned earlier continuing growth in this area is expected. The last word on VCM toxicity must await the final analysis and interpretation of inhalation and ingestion studies which are still not completed. However, VCM levels in PVC packaging are firmly expected to reach such low levels that extraction, if the term still has meaning at these levels, will be at or below the threshold of detection by the most sensitive techniques.

1.8.2 Fire hazards

Fire hazards have been a subject of growing concern in all western communities in recent years. The prevention and control of fires is primarily dependent on the basic design, construction and use of buildings, and the adequacy of the fire precautions enforced. Nevertheless, concern with fire hazards has naturally led to critical assessments of the materials employed in furnishing and building construction, and particularly the plastics and synthetic fabrics increasingly used. Behaviour of materials in fires may be considered under three headings; the tendency to ignite and spread fire, the production of smoke, and the toxic or corrosive effects of the fumes evolved in burning. This is a highly complex subject, and, as in the case of vinyl chloride toxicity, the brief comments here are limited to the possible implications for the pattern of PVC consumption. The polymer is basically of very low flammability, so that PVC compositions, appropriately formulated, will not readily support combustion. This has particularly promoted the use of PVC in certain applications, such as mine belting, and was, for instance, a factor in the choice of PVC chassis runners in the TV application mentioned above. On the other hand, once involved in a fire, sustained by other burning materials, PVC may produce copious smoke. To a large extent, low flammability and smoke evolution from partial combustion are inextricably associated, so that there is no ideal solution. The fumes evolved will contain hydrogen chloride (HCl), an acidic and corrosive gas, and it is this factor which has been of particular concern in hazard studies. The implications for PVC usage in the future depend very much on the emphasis that legislative authorities and consumers place on fire prevention on the one hand, and, on the other hand, on the limitation of smoke and fumes once fires have started. Whilst the former situation could specifically favour PVC, the latter must pose particular problems. A possible approach is to use additives and fillers in PVC compounds which suppress the evolution of smoke and HCl during combustion, whilst retaining the fire retardant properties conferred by the polymer. There are numerous patents claiming additives for this purpose, but it seems likely that effective systems must incur added cost and/or some loss in other properties, particularly in rigid applications. It is interesting to note that B. F. Goodrich has recently introduced such rigid compounds in the USA, although there is, as yet, no market demand in this country.

1.9 THE LONG TERM

At the time of writing, each of the PVC producers in the UK is either constructing, planning or actively considering expansions in capacity, and this in itself speaks for the industry's belief in the continuing expansion of consumption. For the next few decades at least, North Sea oil provides an assured feedstock. In the long term, the oft-repeated warnings of depletion of the world's oil reserves must finally come true, but it is difficult to imagine an industrial society that will not still make and use synthetic polymers. By then, we may be using coal as the primary source of feedstocks, or conceivably have learnt, with the help of enzymes, to synthesise long chain polymers from the constituents of air and water, as the Hevea tree synthesises natural rubber. We shall still require caustic soda, from the electrolysis of brine, inevitably producing chlorine which assuredly will be usefully locked up in PVC.

ACKNOWLEDGEMENTS

The author gratefully acknowledges the permission of ICI (Plastics Division) to publish this contribution, and the assistance of colleagues in the compilation of the statistical data.

REFERENCE

1. Tester, D. A. (1973). In *Developments in PVC Technology*, J. H. L. Henson and A. Whelan, eds. Applied Science Publishers Ltd, London.

Chapter 2

THE MANUFACTURE OF PVC

R. H. Burgess

ICI Plastics Division, Welwyn Garden City, UK

SUMMARY

Over the past few years two factors have dominated developments in the manufacture of PVC. The discovery of a link between exposure to vinyl chloride monomer (VCM) and angiosarcoma, a rare form of liver cancer, in man has caused many modifications to PVC manufacturing plants designed to reduce the exposure to VCM of PVC plant operators, fabricators of articles from PVC, the general public who use PVC especially for food packaging applications and VCM emissions to the atmosphere surrounding PVC plants. Secondly, the continued growth of the overall PVC business has stimulated efforts to reduce the cost of the PVC manufacturing process. This article reviews the developments made in these two areas and pays particular attention to new processes designed to remove unreacted VCM from PVC after the polymerisation process and to the various problems associated with the use of large reactors for suspension polymerisation.

2.1 INTRODUCTION

PVC is one of the major thermoplastic materials with an enviable and continuing growth. The wide diversity of the applications for which it is used virtually guarantees a healthy business at least for the forseeable future. The world market for PVC in recent years and the projected growth over the next few years are shown in Table 1.

There are three major methods of manufacturing PVC; namely by suspension, emulsion and mass polymerisation. Limited quantities of

TABLE 1
WORLD CONSUMPTION OF PVC

Year	1965	1970	1975	1980	1985
Total world sales (M tonne)	3·0	6·0	8·1	13·5[a]	21[a]

[a] Estimated.

speciality polymers are produced by solution polymerisation. The present world capacity for making PVC by these three main processes is shown in Table 2. The proportion of PVC made by each of these three processes now appears to be fairly stable, and it seems unlikely that this will change significantly over the next decade or so.

This steady growth in the PVC business is one of the major spurs for the continuing developments in PVC manufacturing processes but the most important recent developments have stemmed from the discovery that exposure to vinyl chloride monomer (VCM) could cause cancer. Preliminary work on rats by Viola in Italy had indicated the possibility and more extensive animal studies by Maltoni, sponsored by ICI, Solvay, Montedison and Rhone-Poulenc, confirmed that VCM induced tumours, particularly a rare form of liver tumour, angiosarcoma of the liver. However it was the discovery of angiosarcoma cases among workers on B. F. Goodrich's PVC plant at Louisville, Kentucky, USA that alerted the world early in 1974 to the fact that VCM appeared to be a human carcinogen.

TABLE 2
WORLD NAMEPLATE CAPACITY FOR THE MANUFACTURE OF PVC BY THE SUSPENSION, EMULSION AND MASS PROCESSES

Year	*Nameplate capacity (thousand tonnes)*			
	Suspension[a]	*Emulsion*[a]	*Mass*	*Total*
1960	1 430	360	12	1 800
1965	2 900	660	140	3 700
1970	6 200	1 160	340	7 700
1975	9 800	1 700	1 000	12 500

[a] These figures are approximate because, to some extent, emulsion and suspension capacity can be used interchangeably.

Epidemiological studies on workers associated with the manufacture and polymerisation of VCM have subsequently confirmed the link between heavy exposure to VCM and liver cancer. So far a world total of 55–60 cases of angiosarcoma of the liver have been confirmed as occupationally linked to heavy exposure to VCM, all of them in workers involved in manufacturing PVC, generally with many years exposure to very much higher concentrations of VCM than those common in early 1974.

For the past few years VCM toxicity has been a dominant factor for the PVC industry. It has led to significant changes in process operation so as to ensure acceptable plant hygiene standards, environmental protection and low residual VCM in polymer. These measures have consumed a great deal of technical effort, and major capital expenditure programmes have been necessary, all aimed at reducing human exposure to vinyl chloride.

This review considers the development in manufacturing methods in recent years against the background of the need to eliminate the VCM toxicity problem, to reduce the cost of new plants and the need to improve the properties of PVC polymers.

2.2 DEVELOPMENTS DUE TO THE VCM TOXICITY PROBLEM

2.2.1 Protection of PVC Plant Operators

The major problem for PVC manufacturers, on the initial discovery of the link with angiosarcoma, was that of protecting the operators of PVC producing plants from exposure to potentially harmful concentrations of VCM. In most countries co-operation between industry, workers' representatives and government has resulted in rapid progress towards very low concentrations of VCM in the plant atmosphere. For example the British Code of Practice specifies a maximum time weighted average (TWA) of 10 ppm v/v, no excursions beyond 25 ppm, with continuing efforts to produce as low concentrations as possible. This compares with figures (TWA) of 100 ppm common a few years ago and up to 1000 ppm 10–20 years ago. Other government authorities, notably Occupational Safety and Health Administration (OSHA) in the USA who specify < 1 ppm v/v as the average concentration, with no excursions beyond 5 ppm, at first sight are applying more rigid standards. In practice the British Code with its stringent monitoring and excursion requirements is as tight as that proposed by OSHA.[1]

Changes in operating procedure have played a major part in reducing the concentration of VCM in plant atmospheres. Most PVC plant autoclaves are now cleaned automatically (either with high pressure water or with solvents) or recipes are used which do not produce the autoclave build-up which formerly had to be removed between batches. Manual cleaning of autoclaves is known to be a major factor in the exposure of plant operators to VCM.

Interbatch opening of plant autoclaves is another major cause of VCM release into the plant atmosphere. Changes in procedure before opening the autoclave have markedly reduced this emission and some manufacturers have eliminated this source of VCM concentration by operating a non-opening process. A further source of VCM in the plant air is leaks from flanges and valves, especially in the VCM delivery main or on the autoclave itself. Much attention has been paid to eliminating such leaks both by better maintenance and in many cases by replacement with more suitable equipment. Further reductions have been achieved by improved ventilation of the plant either by using forced ventilation (up to 10 air changes per hour) or by opening up the plant to the outside by removing parts of the plant walls.

Finally, residual VCM left in the polymer and, in the suspension polymerisation process, in the waste water, provided another source of VCM. The solution to this problem lies in more efficient VCM removal by stripping after polymerisation. This is dealt with in detail later, but where such a source of VCM exists the problem has been reduced by local ventilation and by increasing the height of drier exhaust stacks which may contain significant quantities of VCM.

All these changes have been very expensive (at least £10 M has been spent in the UK alone), in fact in some cases manufacturers have chosen to close a plant rather than to modify it because of the expense. Where the closure option was chosen, the plants have tended to be small old units containing many small autoclaves and other small pieces of equipment particularly difficult to modify economically. While the techniques necessary to reduce VCM concentrations in the plant atmosphere vary slightly with the type of PVC plant involved, the overall achievement is surprisingly similar whether the plant is based on the suspension, emulsion or the mass polymerisation process.

2.2.2 VCM Analysis

A major factor in the success of this work has been the development of very sophisticated methods of measuring VCM in the working atmosphere and in the solids and liquids from a PVC plant. Both infrared absorption spectroscopy, using the =C—H out-of-plane deformation band at 10·9 μm, and flame ionisation techniques have been widely used. The latter method has been particularly useful when coupled with gas chromatographic separation of organic vapours, making the test specific for VCM.

Colorimetric tests based on the oxidation of VCM with acidic potassium permanganate ($KMnO_4$) or manganese dioxide (MnO_2) to give chlorine or HCl respectively, have been used to give the familiar coloured crystals (Dräger or Gastec) or staining of indicating paper (UEL). Mass spectrometry, using the specific ion technique after separation by a gas chromatograph, has also been used. VCM concentrations down to 0·1 ppm v/v in the working atmosphere are now measured routinely on many PVC and VCM plants. Both infrared and flame ionisation (after gas chromatographic separation) measurements have been linked to computer controlled area sampling, enabling instantaneous and average VCM concentrations to be obtained with little manual effort. This information has been used to identify areas of higher than average VCM concentration for individual attention. A number of portable VCM monitors have been developed (such as the Century Systems Corporation OVA 108 organic vapour analyser) which are used to locate the source of VCM emissions from such things as the flanges and valves on lines containing VCM.

Carbon absorption tubes through which plant air is continuously drawn by small, electrically-safe pumps, have been used for the determination of TWA VCM concentrations, both for area monitoring where a permanent installation is inappropriate and for personal monitoring of the exposure of individual plant operators. The UK Code of Practice requires records of the area monitoring test results to be kept for periods up to 30 years.

Increasing emphasis is now being placed upon the measurement of VCM concentrations round PVC and VCM manufacturing plants, especially at the boundary of the works, to ensure that the general public is not exposed to significant quantities of VCM. Both carbon absorption tubes and colorimetric techniques have been used for this purpose but especially accurate analysis is required in this area (0·1 ppm v/v in air or less).

VCM analysis of the liquid and solid effluents from the plants has also received considerable attention with gas chromatographic analysis followed by flame ionisation being most widely used. Analysis down to 1 ppm w/w is most frequently used but techniques capable of 50 ppb w/w (i.e. 50 parts VCM per 10^9 parts liquid or solid) are readily available both for VCM in water and in PVC.

Most recently VCM concentrations in the PVC sold to customers have been measured routinely by most manufacturers. Although techniques exist to reduce VCM concentrations during the conversion of PVC to fabricated articles, and these are used in some critical areas, increasing pressure is being applied to the PVC manufacturers to reduce the residual VCM in their products to a minimum. PVC articles used to package foodstuffs (e.g. PVC bottles) are already made containing very low VCM concentrations such that the contents of the article always contain very low concentrations of VCM (usually <50 ppb). For these purposes measurements of VCM concentration in dry PVC with an accuracy of 1 ppm are commonly made but head space gas chromatography techniques (using for example a Perkin Elmer F40 or F42 head space analyser) with an accuracy of 0·1 ppm w/w are available.

2.2.3 Removal of Residual VCM

It is a characteristic of the VCM polymerisation kinetics that the reaction slows down markedly at high conversion to PVC, i.e. in the presence of small quantities of VCM, so that it is quite uneconomic to polymerise VCM to completion. In fact most processes stop the polymerisation when 10–20% of VCM on PVC remains.

It has always been the policy of PVC manufacturers to recover as much as possible of this VCM by stripping, i.e. by gasification and subsequent reliquefaction. However the VCM is held tenaciously by the PVC and prior to the VCM toxicity problem most manufacturers operated with a process leaving at least 1% residual VCM on PVC. Most of this residual VCM was then lost during various transfer operations and especially during the drying step necessary with the suspension and emulsion processes. Residual VCM concentrations of 100–1000 ppm w/w were common in freshly dried suspension PVC, and higher concentrations in mass PVC which, of course, has no drying step.

The residual VCM concentration was shown to vary from grade to grade of PVC because of the particle structure of the grades with, as expected, the more porous particles, e.g. those for plasticised applications, giving the lowest values and the dense particles, e.g. those for injection and blow moulding, the highest values. Dried emulsion PVC contained low concentrations of residual VCM (<5 ppm) because of the very small size of the PVC particles and the consequent ease of loss of VCM by diffusion on drying. All dried products continue to lose VCM in storage if stored in paper bags but there is no significant loss in bulk storage.

During 1974 most PVC manufacturers directed their attention to improving their VCM stripping processes with the twin aims of preventing

VCM emissions inside and outside the plant and of reducing the amount of VCM in the PVC supplied to their customers. Figure 1 shows a schematic diagram of the process of removing VCM from a PVC particle, assuming that the PVC particle is suspended in water.

The studies of Berens[2] have shown that in an equilibrium situation containing roughly equal quantities of PVC and water, at least 90% of the VCM present will be in the PVC. Hence the first step in the process of VCM

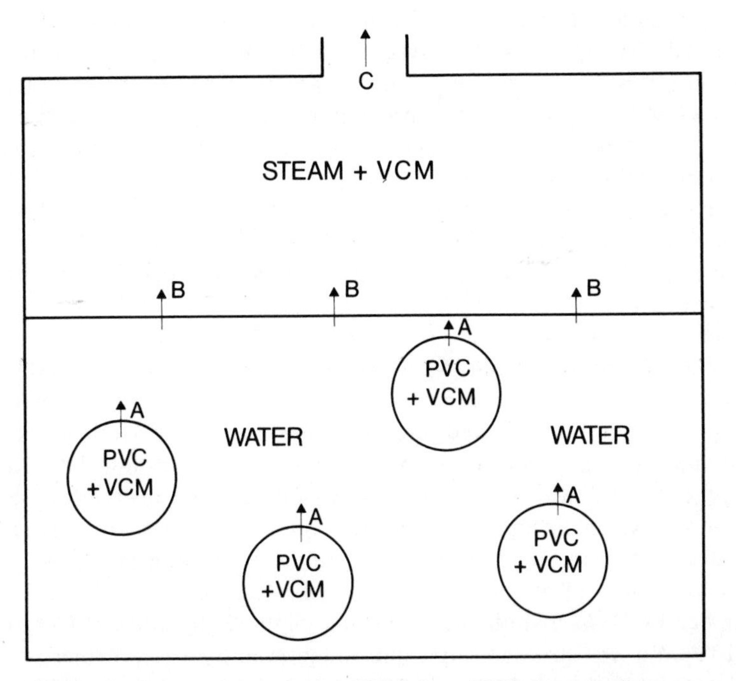

FIG. 1. Process for removing VCM from PVC slurry or latex.

removal is the transfer of VCM from the PVC particle to the water (A in Fig. 1). The rate of this process clearly depends on the diffusion path length, the temperature, and the force driving the VCM from the PVC to the water.

It is already established that some grades of PVC are more difficult to strip than others and for this reason some PVC manufacturers have modified the particle structure of some of their more dense grades to reduce the diffusion path length. The effect of this on VCM release can be demonstrated by reference to Berens' work[3] as can the effect of increasing temperature. In fact most improved stripping processes are based on the use of higher temperatures than were formerly used but care has to be taken to avoid the known effects of too much heat history on the thermal stability of the PVC.[4] Finally, the rate of removal of VCM from the PVC particle is determined by the force driving the VCM into the water which, at a particular temperature and VCM-in-PVC concentration, will depend on the VCM concentration in the water. This, of course, depends on the efficiency of VCM removal from the water (B in Fig. 1) and in most batch stripping processes is a major rate controlling step. In principle VCM can be removed very rapidly from the water by sparging with air[5] but this merely creates a fresh problem of handling a large quantity of air contaminated with VCM.[6] A solution to this problem is to use steam as the carrier gas which can then be condensed to leave relatively pure VCM which can be reliquefied in the normal way.[7] The use of steam as a carrier gas also overcomes the third rate controlling step, that of removing the VCM from the gas space of the vessel (C in Fig. 1). Processes using steam are now widely used to produce suspension PVC with <10 ppm w/w on PVC of residual VCM when dried.

Goodrich has announced a development of this process in which VCM is stripped continuously from PVC slurry. The slurry passes vertically down a column against a counter-current of steam which acts both to increase the temperature of the slurry and carry away the stripped VCM.[8] The column contains a number of trays to control slurry down-flow and steam dispersion. Such a process is inherently more efficient in terms of steam usage than batch stripping. The process is claimed to have little effect on product quality although, as with all continuous processes, good residence time control is important. This process has already been licensed to a number of PVC manufacturers and is claimed to produce PVC with <10 ppm in the stripped slurry and <1 ppm in the dry polymer.

In principle, the elimination of the aqueous phase removes one of the rate controlling steps as indicated by Fig. 1, namely the transfer of VCM from the PVC to the water and from there to the gas space. This improvement is the subject of a number of patents[9] in which VCM is removed from wet cake rather than from slurry. The mass process apparently has the same advantage but the absence of water and, hence, steam to act as a carrier gas means that VCM removal is controlled by the quality of the vacuum achieved. On a large scale it is not easy to obtain a very good vacuum economically and higher residual VCM concentrations result. The addition of some water has been proposed.[10] The use of organic non-solvents for PVC as carrier gases has also been proposed.[11]

By comparison with the stripping of VCM from PVC made by the suspension and mass processes, the removal of residual VCM from emulsion PVC at the latex stage has been the most difficult. The application of heat and vacuum to a latex in the normal batch stripping process inevitably produces large quantities of foam because of the low surface tension of the system. While this problem can be overcome by controlling the rate of off-take of gas or by breaking the foam mechanically, the former process is expensive in equipment time and difficult to control, and the latter is very difficult to achieve at an economic price and coagulum, resulting from the high shear forces used, often results.

These difficulties have been recognised by the US Environmental Protection Agency which is suggesting <2000 ppm residual VCM on PVC after stripping latexes compared with <400 ppm for suspension and mass PVC. A number of processes involving spraying latex into an evacuated chamber have been proposed[12] to overcome these problems. Because of the small size of the PVC particles in a latex the rate of loss of VCM is potentially very high. This is evidenced by the very large reduction in the polymer VCM concentration which occurs on drying. Commonly the concentration falls from 1% to <10 ppm during the spray drying operations.

Most PVC copolymers are made by the suspension polymerisation route and commonly contain 8–16% vinyl acetate to improve melt processing. These polymers tend to have very dense particle structures and this, coupled with their low softening point and consequential low degassing and drying temperatures, means that they are more difficult to degas than homopolymers and commonly retain a higher concentration of residual VCM.

2.2.4 Change of Particle Type

Berens has shown clearly in his work[3] that the rate of loss of VCM from a PVC particle is controlled by the porosity of the particle and has suggested that the effective control is the diffusion path, i.e. the average effective diameter of the particle as measured by the nitrogen absorption surface area. This means that it is desirable from a VCM removal standpoint to increase the porosity of the PVC particles and to make the particles uniformly porous.

The formation of PVC particles in the suspension polymerisation process is largely controlled by the protective colloid used. In recent years cellulose derivatives and partially hydrolysed polyvinyl acetate (PVA) have been most used commercially but the requirement for increased porosity and better uniformity has encouraged the examination of mixed protective colloid systems. Mixtures of cellulosic materials with different substituted groups (methyl, hydroxy ethyl, etc.) have been suggested,[13] as have mixtures of cellulosic and PVA types.[14] Use of short chain surfactants such as the Span and Tween ethyoxylated or esterified sorbitols, mixed with the main protective colloid, have also been proposed.[15]

2.3 DEVELOPMENTS IN THE POLYMERISATION PROCESSES

2.3.1 General Kinetics of the Polymerisation

All of the three major processes for PVC manufacture are based on free radical initiated polymerisation. This can be described by the equations

$$\text{Initiator} \rightarrow 2\text{I}^{\bullet} \quad (1)$$

$$\text{I}^{\bullet} + \text{M} \rightarrow \text{R}^{\bullet} \quad (2)$$

$$\text{R}^{\bullet} + \text{M} \rightarrow \text{R}^{\bullet} \quad (3)$$

$$\text{R}^{\bullet} + \text{M} \rightarrow \text{P} + \text{R}^{\bullet} \quad (4)$$

$$2\text{R}^{\bullet} \rightarrow 2\text{P} \quad (5)$$

where $I^{\bullet}$ are free radicals formed by decomposition of the initiator, $R^{\bullet}$ is a polymer free radical, M is a VCM monomer unit and P is the final polymer molecule. Both the suspension and mass polymerisation processes have essentially the same kinetics but those emulsion processes which use a water soluble initiator, as opposed to the monomer soluble initiator, differ in the detailed kinetics of the initiating step.

A major feature of all three processes is the chain transfer effect of VCM expressed by eqn. (4). The rate of this reaction increases with increasing temperature more rapidly than the chain propagation reaction, eqn. (3). This means that the molecular weight of the PVC made is controlled by the polymerisation temperature used and is little affected by the concentration of initiator. Most of the commercially useful PVC can be made by polymerising VCM at from 50 °C to 75 °C although this temperature range can be extended by using chain transfer agents.

Kinetically, the mass and suspension polymerisation processes are very similar with the rate of reaction being influenced by the initiator used and by the amount of PVC produced. PVC is substantially insoluble in VCM. Polymerisation occurs both in the monomer phase and in the precipitated PVC (which is swollen with VCM) phase. The rate of termination, eqn. (5), is much lower in the polymer-rich phase, because of the well-known Trommsdorf effect, so that the overall rate increases markedly with increasing conversion of VCM to PVC.

Finally, the polymerisation rate falls because of the low rate of VCM diffusion through the highly viscous polymerising system. Very little chain transfer to polymer occurs, so that the PVC produced is substantially free of long chain branching (<1 branch per PVC molecule) and short chain branches arising from molecular rearrangement during polymerisation are also comparatively infrequent (2–5 per 1000 carbon atoms). Consequently the molecular weight distribution of PVC is very narrow and polymer molecular structure is remarkably simple. There is, however, some evidence for both end-group and main-chain unsaturation. This, together with the branched structures and other polymer irregularities, may be responsible for the known differences in the rate of degradation of PVC on subsequent heating. A much more detailed account of the kinetics of VCM polymerisation is given in a recent book.[16]

2.3.2 Suspension Polymerisation

2.3.2.1 *Description of Process*

The process is shown diagrammatically in Fig. 2. A mixture of water, VCM, a free radical initiator and a protective colloid (usually a water soluble polymer such as cellulose or polyvinyl acetate derivatives) is agitated in a jacketed pressure vessel capable of withstanding the pressure generated by liquid VCM at the polymerisation temperature (7–14 bars; 100–200 psi). The temperature of the autoclave and its contents is increased to the polymerising temperature, using hot water in the jacket. The temperature is subsequently controlled by removing the heat of polymerisation by cold water in the jacket. At a predetermined pressure or after a given time, corresponding to 70–90% conversion of VCM to PVC, the residual VCM is removed either in the autoclave or externally as described earlier. The stripped slurry is then dewatered, using a continuous centrifuge, to a wet cake ($\sim$20% water), dried with air and packed either in bags or containers or stored in silos for subsequent sale.

2.3.2.2 *Continuous Polymerisation*

A typical PVC suspension polymerisation plant based on batch autoclave operation would have an autoclave turn-round time in the region of 10–15

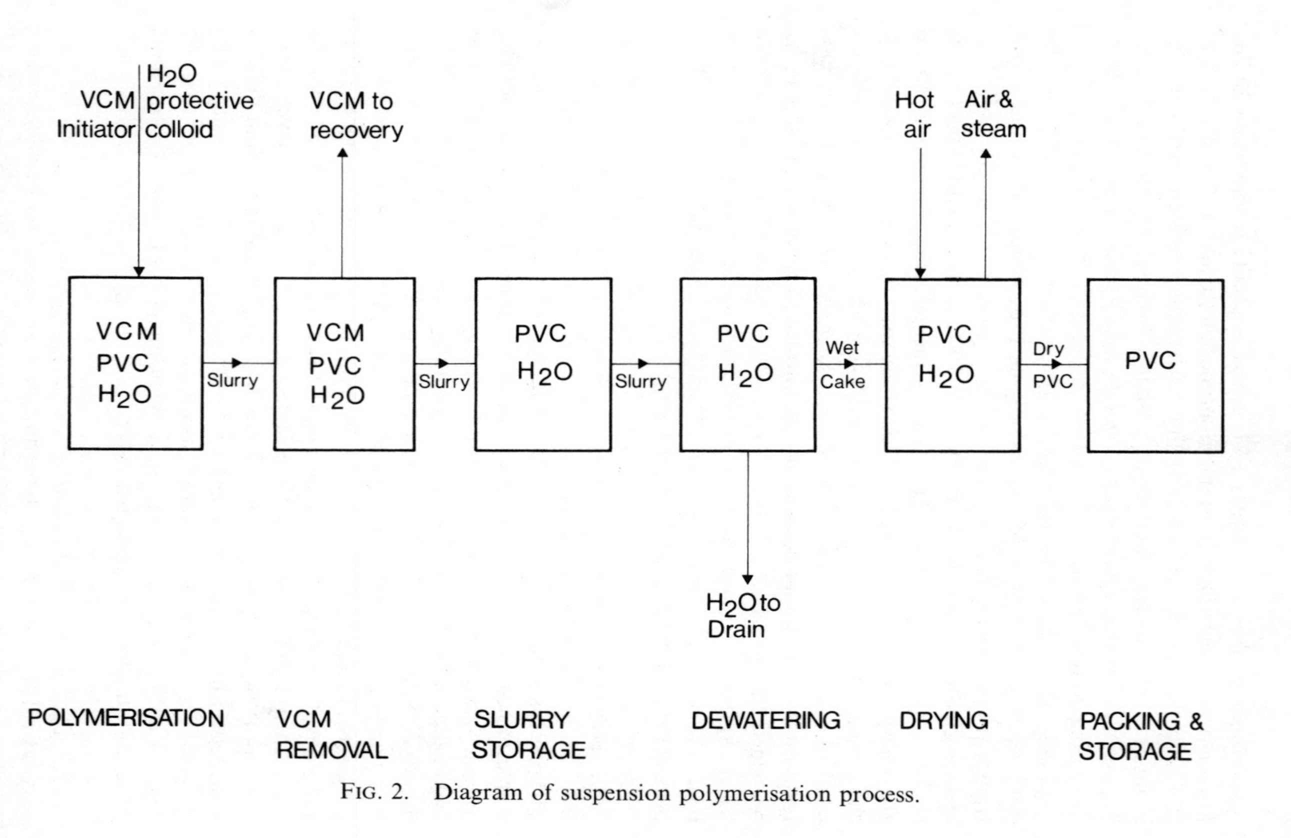

FIG. 2. Diagram of suspension polymerisation process.

hours, depending on the grade and the stripping process used. Of this time as much as 30–40% is taken up by non-productive operations such as charging ingredients, heating up to the polymerisation temperature, residual VCM removal, PVC slurry discharge and autoclave cleaning. In principle all this time could be saved by a continuous process.

Much effort has been expended over the years to this end but there are two major problems to be overcome; build-up on the walls of the polymerisation vessel and control of particle structure. For a continuous process to operate economically a non build-up system is essential; however, most suspension PVC processes produce significant quantities of build-up. Recent developments (see below) suggest that this problem may be soluble. More serious is the particle structure problem. PVC is sold primarily on the characteristics of the actual PVC particles made, i.e. their ability to absorb plasticiser and to melt uniformly. In a batch process the growth of the PVC particles can be controlled satisfactorily but this has not yet been achieved in a continuous polymerisation. Consequently no successful continuous suspension process has yet been evolved.

2.3.2.3 *Size of Polymerisation Autoclave*

The autoclave is the heart of a suspension polymerisation plant and much of the overall cost is tied up with this vessel. When PVC was made at small tonnages, comparatively small autoclaves were used partly because such vessels are easier to produce and operate and partly because the use of a number of vessels reduces the peak services (cooling water, electricity, etc.) demand. As the size of the PVC business has grown, so the size of PVC plants has increased and it has become desirable to use much larger polymerisation vessels. In practice the ratio of autoclave size to plant capacity has not changed greatly over the life of PVC with 1 m^3 (35 ft^3) autoclaves being used for the earlier 1000 tonnes per annum (tpa) plants and 25–200 m^3 (890–7100 ft^3) autoclaves for modern 100 000 tpa plants. Autoclaves of the 40–80 m^3 (1400–2800 ft^3) size are nowadays commonly used by PVC manufacturers but some companies, Shin-Etsu with 135 m^3 (4800 ft^3), and Hüls with 200 m^3 (7100 ft^3) autoclaves, have chosen larger sizes. These decisions are governed by considerations of economics rather than chemistry or engineering. There are a number of engineering problems to overcome when increasing the size of PVC reactors which have been well summarised by Terwiesch of Hüls.[17]

As the size of the vessel increases, so the available cooling capacity, expressed as the ratio m^2 of cooled surface to m^3 of polymerisation volume, inevitably falls. Since the VCM polymerisation process is highly exothermic

(366 kcal per kg VCM polymerised) and good temperature control is important for molecular weight control, this could lead to the need for slower polymerisation rates thus reducing the economic advantage of the large autoclave.

Manufacturers have tackled this problem[17] by increasing the cooling capacity available by reducing the wall thickness, increasing the wall heat transfer coefficient and using chilled water. A number of companies have proposed the use of extra cooling capacity gained by means of an external condenser which removes heat by condensing VCM vapour.

Over the past few years lauroyl peroxide, a traditional VCM polymerisation initiator, has been progressively replaced by free radical initiators which decompose much more rapidly. The intention here is to reduce the initiation rate as the polymerisation proceeds so as to balance the acceleration caused by the reducing termination rate so that the overall rate of heat generation (i.e. polymerisation) is substantially constant. Peroxydicarbonate initiators —— $(ROCOO)_2$ are widely used, but peresters such as *t*-butyl perpivalate, azo initiators such as azo bis(2,4 dimethyl valeronitrile) and acetyl cyclohexyl sulphonyl peroxide are all in routine use. Increasingly, the direct injection of these initiators into the polymerisation vessel is used because of the desirability of avoiding the opening of the autoclave, which leads to increased VCM emissions into plant atmospheres and to an increased autoclave turn-round time.

In suspension polymerisation plants, agitation is normally produced by a single agitator, perhaps associated with baffles. In older plants the agitator was driven by a shaft through the top of the autoclave but as the size of the autoclave increased this shaft became very long and very thick and, of course, very expensive. Bottom entry agitators on a short shaft are now widely used. Baffles of various types have been proposed and in all probability many of these types or none at all are used depending on the agitation required with the particular protective colloid recipes in use.

2.3.2.4 *Use of Computer Control*

The use of computer control of suspension polymerisation plants is now widespread. The advantages claimed[17] include greater safety of operation, the production of a higher proportion of products of the desired high quality, a higher plant utilisation and a reduced manpower requirement. The extent of the computer control varies widely from plant to plant and a system is generally not worth incorporating into an old plant. The use of computers for feed forward control has been advocated but it is likely that most plants use computers as a more rigorous control of the plant process.

2.3.2.5 *Elimination of Autoclave Build-up*

It is a characteristic of suspension polymerisation processes that a deposit of PVC is formed on the walls and other surfaces within the autoclave during each polymerisation. This problem arises because, unlike many other suspension or pearl polymerisations (polystyrene, polymethyl methacrylate), the polymer is insoluble in the monomer and a deposit once formed can only grow. This build-up has to be removed between each batch in order to maintain satisfactory heat transfer to the cooling jacket, and to maintain product quality (by preventing the formation of faults in the PVC, e.g. 'fisheyes' when it is used for plasticised applications).

This cleaning operation was formerly carried out manually but is now carried out either with high pressure water jets[17] or by the use of a suitable solvent.[18] In spite of these improvements the cleaning operation is costly both in terms of cleaning equipment and in extending the batch turn-round time. Considerable effort has been devoted to eliminating the need for the cleaning operation by preventing build-up entirely.

A number of companies, notably Shin-Etsu and Goodrich, claim to be operating build-up free processes commercially but many others are active in the field. Judging from the patent literature, numerous chemical methods are used to suppress build-up but the more successful are based on the principle of an inhibitor for VCM polymerisation which will be active on the surface of the autoclave. Goodrich has patented[19] the use of amine and phenol condensates and ICI[20] the use of high molecular weight amines which are known inhibitors for free radical polymerisations. Similarly Shin-Etsu[21] claim the use of surface active amines such as azo dyestuffs sprayed on the autoclave wall. Apparently these additives work either by absorbing on the autoclave wall or being bound there, and then suppressing any polymerisation which may occur in that area. Even if some build-up is formed by, for example, splashing of polymer on to the wall it will only lightly adhere to the wall and probably be removable by washing with a hose.

Another popular area of experimentation lies in the suppression of a chemical interaction between the autoclave wall (normally stainless steel) and the PVC formed. The patent literature in this area is most confusing with for example both oxidising[22] and reducing[23] agents being suggested. Presumably the efficacy of these treatments depends on the details of the polymerisation process being used.

2.3.2.6 *Product Isolation*

PVC is isolated from its slurry by centrifuging off excess water to form a wet cake. This wet cake is then fed to a hot-air drier which may be a combination flash and long residence time drier or a single-stage, long-residence-time drier (usually of the rotary or fluid bed type). No significant developments have taken place in recent years apart from increasing the size of the units in line with the growth in the suspension PVC market. Great care must be taken with all these driers to avoid polymer hold-up which can lead to overheating and consequential heat stability or black speck problems.

2.3.3 Emulsion Polymerisation

Emulsion polymerisation is used to make general purpose polymers for speciality applications, such as calendered film and thin profile extrusion (where particularly easy processing is required) and battery separators (where a sintering process is used to obtain the desired open structure). It is also used for the production of paste polymers, i.e. polymers which, when suspended in plasticiser, form a relatively low viscosity mix which can be used in the fabrication of gloves and bottle cap inserts, fabric coating, etc. The paste viscosity, a major selling point for these products, is determined to a significant extent by latex particle size and size distribution. Consequently much attention is paid to latex particle size in paste polymer production.

In spite of its comparatively minor share of the total PVC business (see Table 2), there are two major process variants for making PVC latex, batch and continuous polymerisation. The batch process is operated in two quite distinctive ways, microsuspension and conventional emulsion polymerisation.

2.3.3.1 *Microsuspension Polymerisation*

Schematically the process for both microsuspension and batch emulsion polymerisation is as shown in Fig. 3. It is similar to that shown in Fig. 2 for suspension polymerisation except that the polymerisation autoclave is linked to either an homogenising mill or emulsifier/initiator injection equipment. There is no centrifuging step and the whole of the water is removed by evaporation in a spray drier. There is an additional step beyond the drier in paste polymer production in which the polymer produced is ground mechanically using a mill.

Both microsuspension and emulsion processes polymerise VCM in the presence of water and an emulsifying agent (normally a water soluble surfactant such as sodium lauryl sulphate, sodium dodecyl benzene sulphonate, etc.). A *monomer*-soluble initiator is used for the microsuspension process and a microdispersion produced by passing a premix of the polymerisation ingredients through a mechanical mill so that

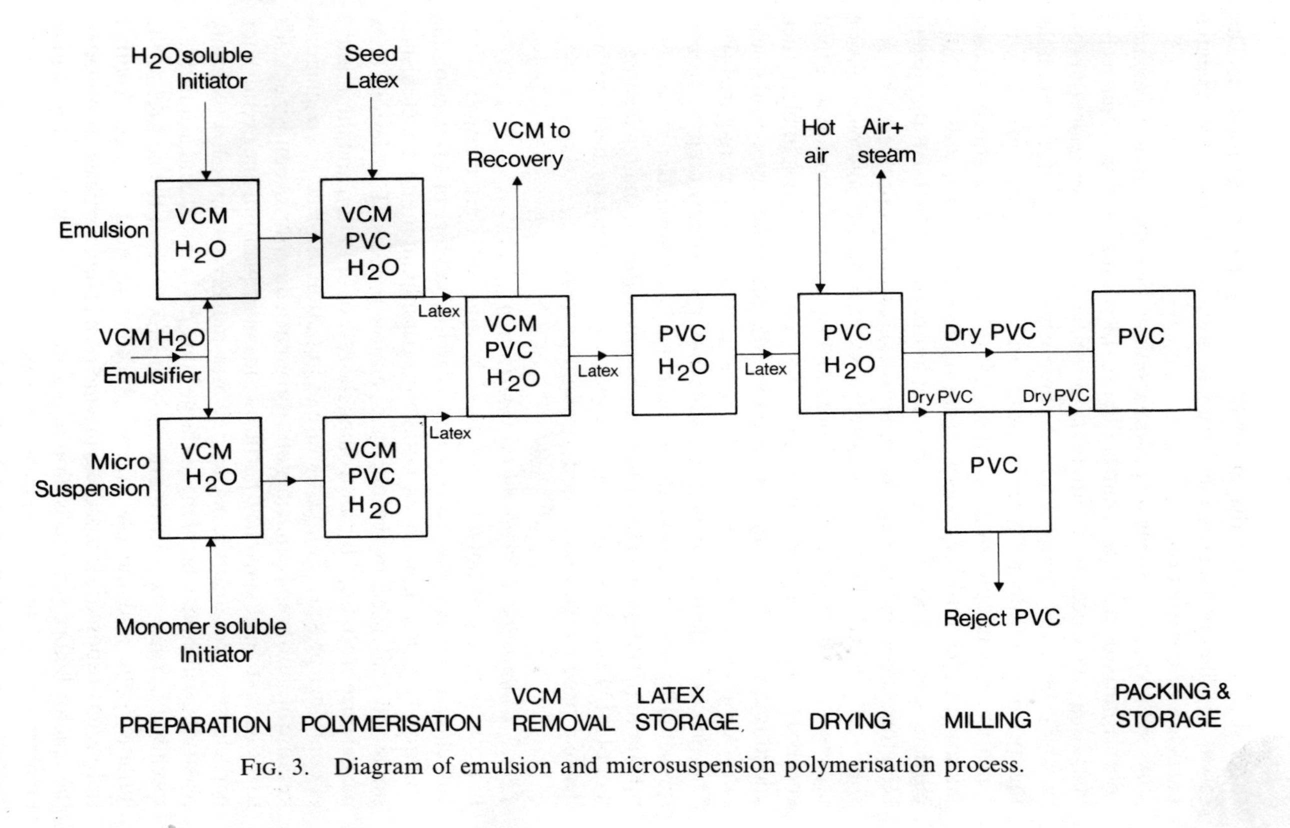

FIG. 3. Diagram of emulsion and microsuspension polymerisation process.

the desirable particle size dispersion (0·1–1 μm particles) is produced. This mixture is then polymerised and the latex retains substantially the initial particle size distribution.

As in the case of suspension polymerisation, the nature of the monomer-soluble initiator used determines the rate of polymerisation. Over recent years there have been many attempts to replace the acyl peroxides (lauroyl peroxide, etc.) with more rapidly decomposing initiators of, for example, the peroxydicarbonate type in order to increase plant productivity. Unfortunately initiators of this type tend to produce some aqueous phase initiation which produces fine particles thus affecting latex stability, latex particle size distribution and product properties. Much of the patent literature is concerned with attempts to overcome this problem.[24]

Recently Rhone-Poulenc disclosed a variant on the microsuspension process[25] in which a seed latex is prepared by microsuspension and then checked for latex particle size and distribution. This seed, if suitable, is then added to a further quantity of water, VCM, and surfactant, and the seed particles grow under polymerisation conditions to the final required size. This process is claimed to make possible the use of more productive recipes (presumably using more rapidly decomposing initiators), to reduce build-up and to permit polymerisation to high conversion. Its main advantage is its potential for the production of a more reproducible particle size, and hence a product with more uniform final properties.

2.3.3.2 *Conventional Emulsion Polymerisation*

In normal emulsion polymerisation VCM and water are charged to the autoclave and polymerisation takes place in the presence of a *water*-soluble initiator (e.g. ammonium or potassium persulphate) and an emulsifier similar to that used in the microsuspension process. The desired latex particle size is obtained by controlling the rate of initiation and the amount of emulsifier present. Too fast a rate of initiation can lead to coagulation because of shortage of emulsifier and too great a quantity of emulsifier leads to the formation of many new particles and consequently a fine overall latex particle size. Because the paste polymer market requires low viscosity plastisols, which in general means large latex particle size and a controlled spread of particle size, much attention has focused on the control of the initiation rate and the use of low emulsifier content recipes. Another technique proposed[26] is a process in which seed latex particles are added to the reactor before polymerisation and grown to the required size. Either uniform or non-uniform particles can be produced, depending on the quantity and size of the seed particles added.

2.3.3.3 *Continuous Polymerisation*

In the continuous emulsion polymerisation of VCM, a premix of VCM, water and emulsifier is fed with a solution of initiator in water continuously to a fairly tall autoclave with a length:diameter ratio of 3:1–4:1. The vessel is agitated so that an excellent dispersion of the two mixes is obtained in the top of the vessel with just sufficient agitation in the remainder of the vessel to ensure a uniform system and good heat transfer. A PVC latex containing substantial quantities of unreacted VCM is removed continuously from the bottom of the reactor at a rate balancing the rate of addition of reactants.

The average residence time is long (several hours) and the agitation is sufficient for the process to be described as a continuously stirred tank reactor with the consequent wide range of polymer residence time. High emulsifier concentrations are used to reduce the amount of build-up formed. This fact and the poor residence time control make the process less flexible than the batch process for the control of latex particle size and hence of the properties of paste polymers particularly. Nonetheless, the process is capable of producing high quality paste polymers. However, in recent years most of the new capacity to make polymers by emulsion polymerisation has been based on batch polymerisation.

2.3.3.4 *Polymer Isolation*

In emulsion polymerisation all the water has to be removed by the drying operation and this makes the drier comparatively more expensive than that in the suspension polymerisation process. Much attention is paid in the polymerisation stage to producing latices of as high solids content as possible. For example, the water content of a 45% solids latex is only just over half that of a 30% solids latex when expressed in terms of water per tonne PVC produced. Hence much less energy is needed to dry the higher solids content latex.

Two main types of drier are in use and both depend on the production of a fine spray of PVC latex which is then dried with a large quantity of air before the particles impact the sides of the drying chamber. In a wheel spray drier the latex is atomised at the periphery of a wheel rotating at high speed into which latex is fed. In nozzle spray driers fine latex particles are produced by forcing latex through a small hole at high pressure (single fluid type) or by atomising latex particles with a cocurrent of compressed air (two fluid type). The choice of the type of nozzle drier depends on the type of PVC particle required.

The drying conditions used significantly affect the properties of the dry polymer made both in terms of subsequent paste viscosity and in terms of the types of particle produced. At the extremes the particles can be comparatively coarse, producing a free flowing powder for use in calendering and extrusion, or fine, with poor powder flow characteristics but excellent rapid dispersion properties when suspended in plasticiser for use as a paste. In recent years the size of spray drying units has increased in line with the growth of the emulsion polymer market.

Milling using fairly conventional grinding equipment is widely practised. Its use reduces the mixing time required to make a usable PVC paste and removes hard particles; these would otherwise disfigure the finally fabricated articles especially when spreading techniques, as used, for example, in vinyl wallpaper manufacture, are used.

2.3.4 Mass Polymerisation

In mass polymerisation VCM is polymerised to PVC in the absence of water. The only commercially successful mass polymerisation process is that developed by Rhone-Poulenc over 10 years ago. The process is represented schematically in Fig. 4 and has been described in some detail recently.[27]

2.3.4.1 *Pre-polymerisation*

A very porous PVC seed particle is produced in a pre-polymerisation vessel equipped with very high speed agitation. It is believed that very fast rates of initiation are used in this vessel such that the polymerisation is substantially complete at 7–10% conversion before transfer to the post-polymerisation vessel. A number of post-polymerisers can be fed from one pre-polymeriser.

It is the use of this pre-polymerisation vessel which is the key to the successful development of the mass process since the final PVC particle type is substantially determined by the nature of the seed. For example the tip speed of the agitator in the pre-polymerisation vessel controls the particle size of the seed and hence the final product. To obtain the final desired 150 μm particle size a tip speed of 5 m/s (16·4 ft/s) is used. For paste extender grades (particle size ~70 μm) 11–12 m/s (36–39 ft/s) tip speeds are required.

The plasticiser absorption properties of the final PVC are similarly affected by the temperature of pre-polymerisation with a temperature of 62 °C being used for porous grades for plasticised applications and 75 °C for rigid extrusion grades. The desired final polymer molecular weight is obtained by performing the post-polymerisation at the temperature normally used to produce that molecular weight.

The pre-polymerisation agitation requires a very large power input but

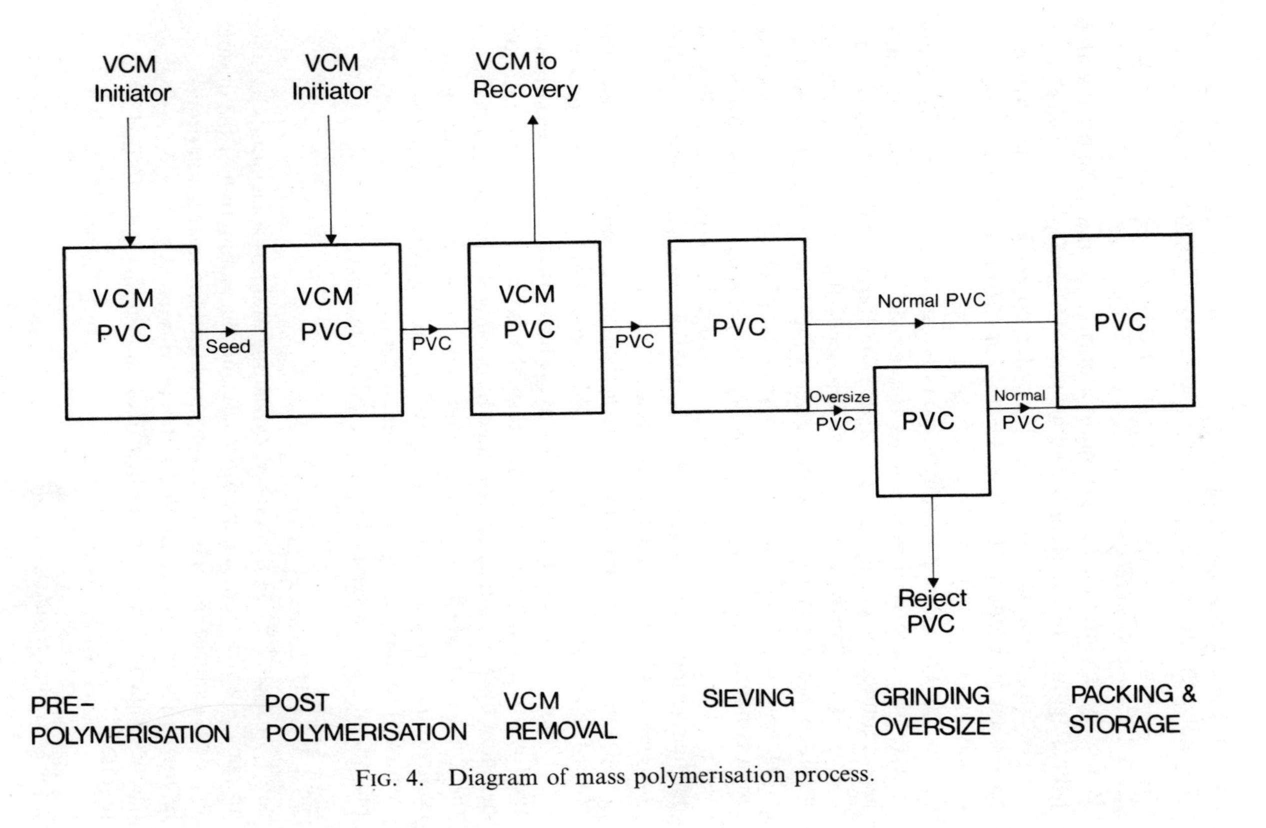

FIG. 4. Diagram of mass polymerisation process.

this has not prevented scale-up of the process from the early 16 m^3 (570 ft^3) post-polymerisation vessels to the present 50 m^3 (1800 ft^3) vessels.

2.3.4.2 *Post-polymerisation*

Post-polymerisation is carried out in a horizontal autoclave fitted with a helical agitator of the ribbon blender type. Up to 5 post-polymerisation vessels can be supplied from one pre-polymeriser. Two major problems have been encountered in the scale-up from 16 m^3 to 50 m^3 autoclaves, namely agitation and heat transfer.

The agitator on the post-polymerisation vessel requires high torque, but nevertheless successful power units for 50 m^3 (1800 ft^3) autoclaves have been produced and a design is available for 100 m^3 (3500 ft^3) autoclaves.

As in the suspension polymerisation process, the available heat transfer surface expressed as a function of the heat generated, i.e. PVC formed, falls with the size of the autoclave so that beyond 16 m^3 (570 ft^3) it is necessary to use a condenser to remove reaction heat even though the agitator is also cooled. The porosity of the particles used and hence their suitability for plasticised and rigid applications falls with increasing conversion so that conversions above 80 % are rarely used.

Build-up is formed in the post-polymerisation but can be reduced by the correct use of polymerisation recipe. Monomer purity is known to affect build-up as is the initiator type used. It is known that cool surfaces reduce the amount of build-up so that it is normal practice to remove the reaction heat first by cooling the agitator, then by cooling the jacket and only using the condenser when the agitator and jacket cooling are insufficient to control the polymerisation temperature at that desired. The autoclave is cleaned automatically by high pressure water once or twice each week. The pre-polymerisation vessel does not require cleaning.

2.3.4.3 *VCM Removal*

Most unreacted VCM is removed by venting and post-evacuation but the absence of water from the mass process and the elimination of the drying step does increase the difficulty, compared with the suspension process, of removing the last traces of VCM. This is discussed earlier. Nevertheless Rhone-Poulenc claims to know how to obtain 1 ppm residual VCM if this is required.

2.3.4.4 *Second Grade Material*

Unlike the suspension polymerisation process the mass process produces significant quantities of oversize material which has to be reground before use as second grade material in non-critical applications. Here the increase in autoclave size has significantly reduced the magnitude of the problem, for example, reducing the oversize from 5–10 % in small autoclaves to <5 % in large autoclaves.

2.3.5 Gas Phase Polymerisation

A number of companies have shown interest in the polymerisation of VCM at subsaturation pressures of VCM. The process consists of the continuous addition of VCM to a polymerising system containing PVC and free radical initiators, such that the pressure is always below the saturation vapour pressure of VCM at that temperature. VCM is absorbed by the PVC particles and polymerisation occurs.

An advantage of the process is that very fast polymerisation rates are possible because the whole of the polymerisation proceeds at the peak rate owing to the Trommsdorf effect at these VCM concentrations (70–80 PVC/30–20 VCM). A further advantage is that build-up of PVC on the walls of the polymerisation vessel is very slight.

The process is in many ways similar to the conventional mass process in that seed PVC particles are used and the properties of these seed particles influence the final product produced. Processes similar to the conventional suspension[28] and mass processes[29] have been examined but the most interesting is that in which a fluid bed reactor using PVC particles as the bed and VCM as the fluidising gas has been used.[30] A particular advantage of this process is that it is possible to remove the reaction heat using the latent heat of vaporisation of VCM coupled with external reliquefaction.[31] This process has not been commercialised, presumably because of difficulty in obtaining the correct particle structure for traditional PVC applications.

ACKNOWLEDGEMENT

The author gratefully acknowledges the permission of ICI (Plastics Division) to publish this paper.

REFERENCES

1. BARNES, A. W. *BPF/PRI Joint Conference: Vinyl Chloride and Safety at Work*, 25 May 1975.
2. BERENS, A. R. *Polymer Preprints*, **15**(2), 197, (1974).

3. BERENS, A. R. *Polymer Preprints*, **15**(2) 203 (1974).
4. NASS, L. I. *Encyclopaedia of PVC*, Marcel Dekker New York, 1976.
5. Mitsui, Japanese Patent 1028890.
6. Mitsui, Japanese Patent 1053588.
7. Solvay, Belgian Patent 793505. Hoechst, German Patent 2429777. Hüls, Belgian Patent 832866. Tekkosha, Japanese Patent 1055390.
8. Goodrich, Belgian Patent 843624. Hoechst, German Patent 2521780.
9. Shin-Etsu, Belgian Patent 831964. Hüls, Belgian Patent 830689. Nara, Japanese Patent 1089587.
10. Rhone-Poulenc, German Patent 2612748.
11. Electro Chem. Ind., Japanese Patent 1017950. Sun Arrow Chem. KK, Japanese Patent 1059986. Unilever, German Patent 2612096.
12. Hoechst, Belgian Patent 831744. Tenneco, German Patent 2608078.
13. BASF, German Patent 2541372. Shin-Etsu, Belgian Patent 779861.
14. Hüls, German Patent 2528950.
15. Firestone, US Patent 4000355.
16. JOHNSTON, C. W. *Encyclopaedia of PVC*, Marcel Dekker New York, L. I. Nass, ed., 1976.
17. TERWIESCH, B., *Hydrocarbon Processing*, November 1976, 117.
18. IAMMARTINO, N. R. *Chemical Engineering*, 24 November 1975, 25. *SIR Modern Plastics International*, **5**(7), 1975.
19. Goodrich, Belgian Patent 833558. Goodrich, Belgian Patent 840600.
20. ICI, British Patent 1444360. ICI, British Patent 1439339.
21. Shin-Etsu, British Patent 1291145.
22. Shin-Etsu, British Patent 1373286.
23. Wacker, French Patent 2185678.
24. ICI, British Patent 978875.
25. Rhone-Poulenc, *Informations Chemie Special Export*, 1975, 87.
26. Montecatini Edison, British Patent 1120410.
27. CHATELAIN, J. Paper given at *Second International Symposium on PVC*, Lyons, July 1976.
28. Solvay, Belgian Patent 734046. Firestone, Belgian Patent 729547.
29. Solvay, Belgian Patent 686088.
30. Montecatini, Belgian Patent 757214. Philips, US Patent 3622553.
31. Solvay, Belgian Patent 762552.

Chapter 3

TOXICITY AND VINYL CHLORIDE

H. M. Clayton
British Industrial Plastics Ltd, Darlington, UK

SUMMARY

After nearly 50 *years of commercial production of PVC, vinyl chloride monomer was found to be a human carcinogen. The following chapter describes how contact with vinyl chloride both in the work place and in consumer applications has been reduced to a level where no major hazard exists to health. Data are presented showing how great the progress has been with the problem since it was first recognised at the end of* 1973. *The technology used to bring about such results is reviewed. Further progress will be significantly less dramatic. Although the health precautions taken have led to an increase in the manufacturing costs of PVC, it remains an essential component of present-day civilisation. There is, in fact, an increased awareness of its remarkable versatility.*

3.1 INTRODUCTION

Some future historian of the chemical industry will, in retrospect, point to the vinyl chloride (VCM) problem as a landmark in the progress of its offshoot, the plastics industry. Long periods of exposure to a substance hitherto regarded as relatively medically safe led to the appearance of a number of occupationally related diseases, including a rare form of incurable liver cancer—angiosarcoma of the liver.

The intensive investigations set in train by this discovery are by no means complete—nor is the development of improved manufacturing methods and procedures to reduce still further the traces of VCM escaping into the atmosphere (in polymerisation and fabrication plants) or, migrating into foodstuffs (albeit in minute quantities). Medical research programmes continue into improved and early diagnosis of VCM-related diseases as well as into understanding the mechanisms of its various effects.

In spite of the fact that any report on the problem must, to some extent, be regarded as interim, the PVC industry can point to considerable progress made during the last three years in solving the various technical problems encountered in reducing any possible hazard to a socially acceptable level.

The author has been actively involved in one of the UK Working Parties dealing with the VCM problem and has first-hand experience of the extensive co-operation among companies (both national and international), trade unions and government agencies in order to bring about a speedy solution to the various problems. In spite of the differences of detail, broadly similar approaches to problem resolution have been employed around the world. There are useful accounts of the problem seen through German eyes,[37] and through American eyes:[11] the latter includes a chronology of key events.

3.1.1 Background to the Problem

Large scale production of PVC resins by the polymerisation of VCM started in the 1930s. Although the hazards of fire, explosion and narcosis were well recognised, there was no evidence to suggest that VCM constituted a health hazard; indeed it had been suggested that it would have been adopted as an anaesthetic, but for the fire and explosion hazard.

The first work suggesting a potential for carcinogenicity came from studies carried out by Viola and co-workers in an attempt to reproduce acro-osteolysis in rats.[1] (Suggestions that Russian workers had observed liver damage amongst polymerisation workers as early as 1949 are unfounded.[3] The study related, in fact, to 48 workers compounding PVC with chlorinated diphenyl). Acro-osteolysis, Raynaud's Syndrome and scleroderma had been observed amongst some PVC polymerisation workers, particularly those involved in the manual cleaning of deposits from autoclave interior walls. The act of removing deposits of encrusted PVC resin releases unreacted VCM and it is this VCM which presumably causes diseases such as acro-osteolysis—a vascular disturbance.

The clinical findings may be, for example:

1. Increased sensitivity to cold,
2. Pins and needles and other paraesthesia, and
3. Changes in the colour of the skin of the digits.

These are the signs of a disturbance known as Raynaud's Syndrome. Another feature of the acro-osteolysis may be changes in the bone pattern of the terminal phalanges, often accompanied by deformation of the nails.

Scleroderma is the medical term for the thickening of the skin which can occur as a result of contact with VCM, and it can lead to gross deformation of, for example, the thumbs.

In studies reported in 1971, Viola showed that exposure of rats to atmospheric concentration as high as 30 000 ppm (parts per million) failed to produce acro-osteolysis. However, malignant tumours were found.

The level of exposure of the rats was so high that its significance to human beings was not clear and further work was necessary. A consortium of European interests, including ICI Limited (UK), Rhone-Poulenc-Polymeres (France), Montedison (Italy) and Solvay (Belgium) commissioned Professor Cesare Maltoni to carry out further studies. Cancers were confirmed in animals at exposures as low as 250 ppm, including angiosarcoma of the liver.[2]

Maltoni's findings led to an epidemiological survey being conducted in the USA on men who had previously worked in a PVC polymerisation plant (B. F. Goodrich, Louisville, Kentucky). This study revealed a cluster of cases of the exceedingly rare form of cancer, angiosarcoma of the liver. The results of this study group (by Creech and Johnson) were transmitted to the National Institute of Occupational Safety & Health (USA) on 22 January 1974, the Department of the Employment (UK) on 23 January 1974 and the Acting Chief Employment Medical Adviser (UK) on the next day. Other national governments were notified during the same period.

In the UK on 29 January 1974, ICI issued a press statement announcing that government departments, the TUC, its own workers and customers, were being informed of the facts available so far concerning the urgent investigations into the cause of death of a seventy-year old retired autoclave worker. It was subsequently confirmed that the worker, who had died in 1972, was suffering from angiosarcoma of the liver. During his working life he had undergone twenty years exposure to VCM.

In February 1974 the Chemical Industries Association in the UK formed a Vinyl Chloride Committee to study the whole problem and institute necessary programmes of work through a series of specialised working parties. Similar arrangements were made in other countries of the world. In Europe international co-operation was ensured through CEFIC (Conseil Europeen de Federations de l'Industrie Chimique) which formed its own Vinyl Chloride Committee with national representatives taking decisions on behalf of their colleagues.

3.1.2 Extent of the Vinyl Chloride Problem

At the end of 1976 the number of notified angiosarcoma deaths was fifty-five. The distribution of deaths is shown in Table 1.

The average length of exposure before death was 15–20 years. Almost without exception the deceased had worked on polymerisation plants and had been engaged in the manual cleaning of autoclaves during part of their service. Certain plants show a clustering of deaths.[6]

TABLE 1
NUMBER OF NOTIFIED DEATHS FROM ANGIOSARCOMA (UP TO THE END OF 1976)

Country	*Number of deaths*
Belgium	1
Norway	1
UK	2
Italy	3
Sweden	3
Eastern Europe	4
France	6
Germany	7
Canada	10
USA	18

Crampton[6] and other authors[4] describe other disorders caused by VCM. These include acro-osteolysis (approximately sixty-five cases in the world—all autoclave cleaners), associated Raynaud's Syndrome and scleroderma, and liver dysfunctions. This latter group of disorders has sometimes been referred to as 'PVC Disease'.[7]

Most of the workers suffering from this group of disorders were polymerisation workers and particularly those manually cleaning autoclaves.

Whilst some disagreement exists on the severity of the effects of the other groups of disorders, angiosarcoma of the liver is invariably fatal.

Studies on the carcinogenicity of VCM, whether by inhalation or ingestion, will not be complete for some considerable time. Until a medically proved acceptable exposure level is established, the object of much of the work has, during the last three years, been to reduce possible exposure to VCM to as near zero as possible.

In the meantime, epidemiological surveys of polymerisation workers have been carried out in the UK, the USA and elsewhere to establish the possible extent of the problem.

Other work has been directed at improved diagnostic methods to give advanced warning of problems.[4,8]

Yet further work has been aimed at understanding the mechanism of the way in which VCM causes angiosarcoma of the liver. Tentative hypotheses have been advanced involving the role of metabolites of VCM, such as chlorethylene oxide and modification of DNA molecules.[35]

Although no final conclusions can yet be drawn from the medical work carried out, some authorities have attempted an assessment of the risk from VCM in the light of recent performance. The results are encouraging.[9,10]

Whilst there must inevitably be an element of speculation in assessing all medical work, the results of work by the PVC industry are demonstrable. These will be described in the following pages.

3.1.3 Sources of Interface Between Human Beings and VCM

To understand the problem and the progress made in containing it, a brief account of the processes for the manufacture of PVC resins, compounds and finished articles is necessary.

VCM is a gas boiling at $-13{\cdot}5\,°C$. It is therefore necessary to polymerise it under pressure. Commercial resins are made generally by polymerising 70–90% of the VCM. This leaves a quantity of unreacted gas within the autoclaves as well as within the particles of the PVC resins. Complete reaction of the monomer is not feasible. If made by the suspension or emulsion processes (accounting between them for the majority of world PVC production) it is necessary to separate the water present, for example by centrifuging and drying. All these processes lead to the release of further quantities of VCM. Finally, the resin itself will still contain small quantities of unreacted monomer, particularly those resin types with dense glassy particles used for direct powder blend processing, e.g. in the extrusion of water pipe or extrusion blow moulding of bottles.

Subsequent processing operations involving the use of heat will lead to further releases of VCM. Small quantities remaining in the finished articles will be released by slow diffusion. In the case of packaging materials, very small quantities of VCM may be transferred into foodstuffs and beverages. There is also the possibility of transfer of very small traces of VCM into drinking water supplied through PVC pipes.

3.2 DEVELOPMENTS IN REDUCING CONTACT WITH VINYL CHLORIDE

The preceding section identified the possibilities of exposure to VCM in the following areas:

1. Polymerisation and associated operations,
2. Storage and handling of PVC resins,
3. Hot processing of PVC compounds,
4. Fabrication of finished products, and
5. Diffusion from finished products, including migration into foodstuffs and drinking water.

Before considering the developments that have taken and still are taking place, some estimate of exposure levels is appropriate. Before the realisation that excessive exposure to VCM could lead to crippling or even fatal illnesses, little reliable data on atmospheric VCM levels around polymerisation plants was available. Table 2 gives such information, and is based on more recent measurements and earlier estimates.

TABLE 2
VCM LEVELS AROUND POLYMERISATION PLANTS

Year	*Typical average level (ppm)*
1945–1955	500–1 000
1955–1960	400–500
1960–1970	300–400
Mid–1973	150
Mid–1975	5
End 1976	2–5

Levels encountered in the fabrication and end use sectors are given in Table 3.

TABLE 3
TYPICAL RECENT VCM LEVELS (PPM)

Location	*January* 1974	*July* 1975	*December* 1976
In plant atmospheres	2–15	<2	<1
In PVC bottles	~50	~2	<1
In beverages	0·1	0·01	<0·005

Estimates of the intake of VCM have also been made;[4, 5] these are shown in Table 4.

As expected, the figures show that the intake of VCM by polymerisation plant workers was by far the most significant. As will be shown, many of the steps taken to reduce the exposure of polymerisation workers to VCM led to improvements in downstream activities.

TABLE 4
ESTIMATES OF DAILY VCM INTAKE

Type of person	*Daily VCM intake (g/kg body weight)*
Polymerisation worker at	
1 000 ppm	0·36
500 ppm	0·18
5 ppm	0·001 8
2 ppm	0·000 7
Fabrication plant operator	
1 ppm	0·000 4
European citizen—through ingestion of food	0·000 000 002–0·000 000 008

3.2.1 Polymerisation Plant Developments

The level of VCM permitted in the atmosphere about polymerisation plants, the absolute quantities which may be emitted into the atmosphere, the quantities which may be present in waste water and in some cases the quantities remaining in PVC resins, are the subject of various compulsory national regulations or voluntary codes of practice and hygiene standards. They vary in detail and in the methods of calculating certain required numerical data, in particular average and peak personnel exposure results. These apparent differences obscure the general similar level of achievement in most countries of the world.

Typical average levels of atmospheric VCM about PVC polymerisation plants are now 2–5 ppm; this has been achieved by improving the operating procedures in a number of ways:

1. More efficient degassing of the PVC resin, either in the polymerisation autoclave or downstream from the autoclave, resulting in lower losses at the drier stage. Much effort has been

expended on the development of improved degassing systems and a number of patents have appeared.[19,23,31,32,33] The B.F. Goodrich process for counter-current stripping of resin slurry has been licenced to a number of other PVC resin producers.

2. Development of automated high pressure washing methods for autoclaves so as to avoid the necessity of entry and manual cleaning.
3. Development of autoclave surface-treatment systems which reduce the tendency of crusts or skins to form, so reducing the amount of cleaning necessary.[12,13,18,20,21,24] The Shin-Etsu Company of Japan and the B.F. Goodrich Company of the USA have licenced their technology to a number of other PVC resin producers. Other companies have reported the use of additives in polymerisation recipes which reduce encrustation.[14,22,34]
4. Remote operation of autoclaves which may often be controlled by a computer. This technique will grow in importance as a result of recent spectacular explosions, such as that at Flixborough.
5. The use of larger autoclaves, so reducing the number of flanges and valves, etc. for a plant of given output. Such developments were under way before the health problem of VCM was appreciated. The largest reported commercial autoclaves are those operated by Chemische Werke Hüls—200 m^3. B.F. Goodrich operates a plant at Louisville with autoclaves of 62 m^3. The Shin-Etsu Company of Japan uses autoclaves intermediate in size at 130 m^3. All three companies have licenced their larger autoclave technology.
6. Development of recipes and agitation systems yielding more uniform resin particles. Uniformity of resin particle structure is a requirement of the industry, irrespective of the VCM problem. Added impetus has been given to the work since a more uniform particle structure will lead to more uniform degassing. Especial priority has been given to investigations into the production of resins for such applications as water pipe and bottles which normally start from a powder blend. Processing rates are influenced by bulk density and particle shape. The problem is to produce resins with a regular shape and porous structure and at the same time an acceptable bulk density. Some progress has been made towards this objective.
7. Use of respirators by personnel when carrying out operations in areas with unacceptably high atmospheric levels of VCM.
8. Development of automatic sensitive monitoring systems scanning all working areas.

Outstanding problems still exist, notably in reducing losses from certain downstream stages after the autoclaves. The Environmental Protection Agency in the USA has set very stringent requirements which may prove difficult to achieve.[28]

Abramowitz[30] gives a useful account of progress in mass polymer plants.

Some quality problems have inevitably arisen as a result of producers finding themselves forced to bring about rapid improvements with existing equipment or incompletely developed modifications. These considerations, as well as long delivery times of some equipment, have forced companies to use ad hoc interim arrangements. For example, reductions in residual VCM have been effected by the expedient of more severe drying conditions leading to poorer resin colour and stability, poorer flow in hoppers (due to static electricity effects) and reports of different fusion characteristics (probably attributable to subtle surface changes due to the higher drier temperatures).

The cost of modifying some plants has resulted in their closure. Reduction of approximately 10% of plant productivity has occurred in many cases, and production costs have increased by an average of 10–15%. On the other hand, VCM utilisation efficiencies are markedly improved. A further benefit is that the reduced necessity to enter autoclaves for cleaning has improved working conditions.

3.2.2 Reductions of VCM Exposure in Handling, Processing and Fabrication

The preceding section refers to steps taken to reduce residual VCM levels in PVC resins. Some resin particles are more difficult to degas effectively than others. For rigid extrusion, extrusion blow moulding and injection moulding operations (using a powder blend feed rather than a fully compounded granule), dense regular particles are required in order to obtain good powder flow and efficient packing of screws. Such PVC resins retain VCM more tenaciously than those with an open particle structure used for the manufacture of plasticised compounds and plasticised finished products. In early 1974 it was common for such resins to contain 1000–2000 ppm residual VCM. In some cases even more was retained, e.g. in homopolymers made by the mass polymerisation route, and acetate copolymers.

VCM is slowly released from the polymer at room temperature. At elevated temperatures such as those encountered during processing the release is extremely rapid. In the USA, there are reports of VCM accumulating in the free head space of rail cars until the levels reached are potentially capable of explosion. In Europe, fires have occurred at the vents of extruders processing rigid powder blends.[16]

In the early part of 1974, extensive surveys of more than 100 companies were carried out in the UK, in order to establish the levels of VCM encountered during processing and handling operations. The findings[17] are shown in Table 5.

TABLE 5
ATMOSPHERIC VCM LEVELS AROUND PVC PROCESSING AND FABRICATION OPERATION (PPM v/v)

Location	*Range*	*Average*
Polymer warehouses	<2–17	2
Cold mixing equipment	<2–3	<2
High speed mixing[a]	<2–53	3
Banbury type mixer	<2–13	2
Two-roll mill compounding	<2–2	<2
Paste mixing	<2–16	<2
Calendering	<2–3	<2
Fabric and paper coating	<2–5	<2
Pipe extrusion	<2–7	<2
Cable extrusion	<2–4	<2
Injection moulding	<2–4	<2
PVC welding	<2	<2
Vacuum forming	<2	<2
Bottle blowing	<1–2	<1
Extrusion of flexible film	<1–2	<1
Extrusion of rigid sheet	<1–4	<2

[a] Single result of 53 with high acetate copolymer at bin filling point.

Most of the results were of a low order, averaging 2–3 ppm. However, certain unacceptable practices came to light. High speed mixing areas were particularly prone to unacceptably high atmospheric levels. This was not surprising since at temperatures in excess of 110 °C, PVC resins release VCM very rapidly.

Improved operating procedures, such as the installation of localised extraction, the venting of extruders and mixers to the outside of buildings, and good general ventilation in warehouses etc., resulted in levels below 2 ppm by early 1975. By the latter part of 1976 the effects of reducing residual VCM in PVC resins have resulted in levels generally below 1 ppm with many results at 0·1–0·2 ppm; near the limit of detection of the analytical methods used.

3.2.3 Potable Water Supply

Virtually all potable water pipe is produced by extrusion from powder blend. Although some differences in formulation practice have developed with, for instance, Europe favouring the use of solid lead salts as stabilisers and the USA favouring the use of liquid tin compounds, a common feature of the pipe-making industry was the increasing use of high bulk density resins to increase extruder outputs.

As reported earlier, such resins tend to retain residual VCM more tenaciously, and levels of 1000 ppm or more were not uncommon.

As well as the necessity to reduce residual VCM levels in new pipe production, measurements on drinking water passing through existing pipe systems were required.

At one location in the UK, the Chemical Industries Association was able to carry out measurements on drinking water supplied entirely through a rigid PVC pipe network manufactured by one company. The pipes were 102 mm (4 in) and 152 mm (6 in) in diameter. Even in those parts of the network where PVC resins containing 1000 ppm residual VCM had been employed to manufacture the pipe, no VCM could be detected in the drinking water.

Further studies were carried out on houses connected by small bore PVC and chlorinated PVC pipe. Here, a more unfavourable surface/volume ratio could be expected to increase the possibility of extraction of VCM; however, only at one location could any VCM be detected. This was in a home built some nine or ten months previously and unoccupied since completion. In the small volume of water which had stood in contact with the PVC pipe, 0·015 ppm of VCM was detected.[25]

The above results were extremely encouraging. Nevertheless, three lines of attack were used to reduce VCM levels in rigid PVC pipe:

1. Increased use of hot (i.e. $>110\,°C$), high speed mixing, combined with aeration, in the preparation of the powder blends. Some companies had originally only mixed to relatively low temperatures, i.e. 70–80 °C.
2. Improved degassing techniques during PVC resin production. Typical average residual VCM levels for pipe resins are now in the 5–50 ppm range.

3. Development of suspension resins of increased porosity. A judicious selection of suspending agent combinations, or other appropriate formula modifications in the case of mass polymer, can produce resins with regular particle shape, significantly increased porosity and only a marginal sacrifice in bulk density. Further developments can be expected by this approach.

A combination of the above approaches has led to significant reductions in the residual VCM levels in rigid PVC potable water pipe and some companies are now achieving average levels of less than 1 ppm.

Both the Ethyl Corporation in the USA and ICI in the UK have determined the diffusion coefficient of VCM from rigid PVC pipe. The results have been used to predict VCM levels in drinking water supplied to typical households, in the USA and the UK, using pipe with 1 ppm residual VCM.[26] For the UK, the level of VCM in the water is predicted to lie between 0·006 ppb and 0·022 ppb (b = 1000 million). In the USA the predicted values are 0·01–0·14 ppb. Not surprisingly, the use of liquid tin stabilisers in the USA in place of solid lead salts in the UK gives a higher coefficient of diffusion.

Since most water consumed is used in cooking or in the preparation of hot beverages, with attendant further loss of VCM, the average levels in consumed water are decreased even further.

3.2.4 Food Packaging

Any review of the use of PVC for food packaging must, in the light of the VCM problem, take into account both the political background of legislation and the variations in usage in different countries.

In the USA, the Food and Drugs Administration (FDA) is constrained by precedents set in the interpretation of the so-called Delaney Clause of the Food, Drug and Cosmetic Act of 1958 which states 'No additive shall be deemed to be safe if it is found, after tests which are appropriate for the evaluation of the safety of food additives, to induce cancer in man or animal....'

At the time of its introduction, the above clause was clearly intended to be interpreted in the light of testing, probable consumption levels, safety margins and reasonable certainty about conditions of use. In practice, precedent has established the clause as intending to mean a total ban on any additive suspected of causing cancer under any conditions.

As a result of this, American legislators are not free to make a truly scientific assessment of a possible carcinogenic hazard. Any regulation which falls short of a total prohibition is, in fact, a politically influenced decision.

In this situation the FDA produced a new form of words in a proposed regulation issued in September 1975. The proposals permitted the continuing use of PVC in contact with foodstuffs 'where the potential for migration of vinyl chloride is diminished to the extent that it may not reasonably be expected to become a component of food.' Against this definition the FDA proposed to allow the continuing use of flexible film and hose. Rigid and semi-rigid foils and bottles would have their status of prior sanction withdrawn. Any producer of the latter group of products would then be obliged to demonstrate that they met the FDA criterion.

At the time of writing no decision has been taken on the proposals and it is extremely difficult to predict the outcome.

In the United States the most important outlet for PVC in food packaging is in plasticised films such as meat wrap. On the other hand, the use of rigid PVC is relatively less important and in particular the use of rigid PVC bottles for food packaging has scarcely developed. Therefore, whatever the decision its economic effect on the United States packaging industry will be relatively small.

The EEC is slowly moving towards a unified approach on food packaging legislation, and no matter how much European countries may wish to display their independence of events in the USA, in practice, packaging legislation in the latter country has previously influenced events in Europe. In the case of PVC and VCM, however, the EEC produced a draft directive at the end of 1976 which takes into account the realities of the situation in Europe.[27] Since a 'no effect' level has yet to be established for VCM, the proposed directive takes into account the best practical achievements of European industry in reducing VCM levels in PVC food packaging. Before discussing these achievements, some account of the variations in European food packaging practices is necessary to understand the basis of the proposals. Whilst plasticised PVC films are important for food packaging in Europe, the use of rigid PVC bottles or foils is especially important. Certain countries in Europe show marked variations in the pattern of usage. In France for instance the manufacture of rigid PVC bottles for mineral water is a major industry, accounting for 100 000 tonnes of PVC compound in 1976, whereas in Germany, rigid PVC foils account for the major proportion of rigid PVC used in food packaging.

In the UK a major proportion of rigid PVC bottles is used for the packaging of fruit squashes. Such squashes are mainly consumed by young children, a section of the population for whom there is special concern.

There are also some significant differences in formulating practice in Europe. In France and countries subjected to French influence, most bottles are based on tin-free formulations. Such formulations contain significantly higher levels of other liquid additives to help processing. In Germany particularly, tin stabilisers are preferred.

Whilst there is a significant usage of tin-free bottle formulations in the UK, the majority are based on tin stabilisers. Most rigid PVC foils in Europe are based on homopolymers, but in the UK a significant percentage of production (circa 75 %) is based on acetate copolymers.

TABLE 6

VCM LEVELS IN UK-MANUFACTURED PVC PACKAGING MATERIALS

VCM location	*VCM concentration (ppm)*		
	January 1974	*March* 1975	*December* 1976
Bottles			
Level in polymer at time of use	500–1 000	100–250	15–50
Typical level in powder blend at time of manufacture	50	5	1
Typical level in bottle wall	50	3	<1
Foil			
Level in polymer at time of use	800–1 500	100–250	50–100
Typical level in foil	80	8	3
Flexible film (extrusion blown)			
Level in polymer at time of use	400–1 000	50–150	5–30
Typical level in film	1	1	<1

Table 6 shows the residual VCM levels in UK-manufactured PVC packaging materials during the three-year period January 1974 to December 1976. It will be noted that in the case of flexible film, residual VCM levels were very low, even at the beginning of the period under review. This is explained by the fact that the presence of plasticiser aids the release of VCM during hot processing.

The higher levels of residual VCM in rigid foil compared with bottles is explained by the use of copolymers in the manufacture of the former material. The relatively low softening point of copolymers makes it extremely difficult to achieve low VCM levels either at the resin-making stage or during mixing. For the foreseeable future copolymer foils are likely to contain slightly higher residual VCM levels than bottles or foils based on homopolymers. Since most such foils are either used for the packaging of solid foodstuffs or materials of restricted shelf life, the slightly higher level in the package does not normally lead to higher levels in foodstuffs.

The results shown in Table 6 are also generally typical of the position in Europe as a whole with two provisos:

1. Where homopolymers are used to manufacture rigid foils, residual VCM levels are lower and more in line with UK results for bottles; and
2. Where tin-free stabiliser systems are used in bottle formulations, the presence of extra liquids and low melting point additives assist the displacement of VCM during processing, and residual levels in such bottles will tend to be lower than in bottles based on tin stabilisers.

Some important conclusions can be drawn from Table 6:

1. Flexible film has low residual VCM levels even at the beginning of 1974. Subsequent improvement has only been marginal;
2. Rigid PVC has shown a marked improvement, especially during the period January 1974–March 1975;
3. Since current USA achievement does not differ significantly from that in Europe, the logical course for the Food and Drug Administration would be to accord rigid and semi-rigid PVC the same status as flexible film, i.e. continuing prior sanction.

3.2.5 VCM in Packaged Food

Numerous studies have been carried out on the migration of VCM from the package into the contents. In the UK regular surveys have been carried out on random samples from retail outlets. The results of this large body of continuing work may be summarised as follows:

1. Under ambient conditions rigid PVC packages lose part of their residual VCM to the atmosphere; after a year approximately 30 % has been lost in this way;
2. VCM reaches an equilibrium concentration in the contents of packages after several months;
3. This equilibrium concentration allows typical partition concentrations to be calculated. For foodstuffs commonly packaged in PVC in the UK the concentration of vinyl chloride remaining in the PVC is 200–700 times that present in the contents;

4. Only very small proportions of the VCM in a package may transfer to the contents and for a wide range of foodstuffs the levels of extraction are similar; and
5. Under certain circumstances small numbers of packages may remain on retail shelves for far longer than the recommended shelf life of the contents.

The use of the partition data enables calculations to be made of expected maximum VCM contents of foodstuffs packaged in rigid PVC.

Extensive retail surveys have provided firm support for such predictions. At the end of 1976, VCM levels in foodstuffs packaged in the UK were in the range 0·003–0·005 ppm or less.

Work carried out both in the UK and in Continental Europe has shown that the equilibrium partition concentrations with tin-free bottle formulations such as those typically used in France are not as favourable as those with tin stabilised formulations. For some foodstuffs the ratio of concentration of vinyl chloride in the bottle to that present in the contents may be as low as 100. This disadvantage is, to some extent, offset by the fact that VCM tends to be displaced more readily during hot processing, giving slightly lower initial levels in the bottles.[28]

Gilbert[29] also has suggested that at low initial concentrations in bottles and other rigid containers, extraction of VCM by foodstuffs will be less than expected because of the presence of active sites in the matrix of the PVC which retain VCM.

A further factor influencing the EEC proposals is the state of development of suitable analytical techniques. With some foodstuffs problems of interference exist. Also, even in the hands of highly skilled research analysts, it is very difficult to get agreement between different laboratories.

Finally, the manufacturer of the PVC container or foodstuff must be able to observe trends in the VCM content of his products and take corrective action. For this to be possible, the maximum allowable VCM must be rather more than that at the limit of analytical detection.

Against this background the EEC has produced its draft directive. The main features of the directive are:

1. A limit of 0·050 mg/kg in the foodstuff (i.e. 50 ppb);
2. A limit of 1 mg/kg in the container (i.e. 1 ppm); and
3. In the case of containers made from copolymers, a limit of 5 mg/kg, provided that they are not used for liquid foodstuffs.

It is too early to predict the success of the proposals. However, similar requirements or recommendations already exist in some European countries.

For further information on VCM and PVC packaging the reader is referred to the publications of GECOM (Groupe d'Etude Pour le Conditionnement Moderne). GECOM was founded in France and is an association of PVC producers and the manufacturers of packaging and foodstuffs, including water, oil and wine. It was set up to study the various problems in the packaging field raised by the discovery of the toxicity of VCM. The address of GECOM is 11 Rue Margueritte, Paris 17. One of the GECOM publications gives a very useful account of the problems of analysis of foodstuffs containing vinyl chloride.[38]

3.3 ANALYSIS AND DETECTION OF VCM

The progress described in the preceding pages would not have been possible without significant advances in the methods of detecting VCM. Developments have taken place in three major areas:

1. The monitoring of atmospheric levels of VCM in the vicinity of polymerisation plants;
2. Analysis of VCM, often present only in minute traces, in PVC resins and compounds, food packaging and foodstuffs; and
3. Small portable detectors.

3.3.1 Monitoring of Atmospheric Levels

Prior to 1974 the measurement of VCM was a time-consuming process and the methods required development so that they were capable of yielding accurate results at great speed and were suitable for the sequential scanning of working areas on polymerisation plants.

Two methods of analysis have gained particular favour for the task:

1. Gas chromatography combined with a flame ionisation detector; and
2. Infrared absorption.

In the first method samples of air are drawn through a suitable column which separates VCM from other possible airborne contaminants. The VCM fraction is burnt in the presence of pure hydrogen, forming ions which are then measured by an electrical technique.

This method is very sensitive and can measure fractions of 1 ppm of VCM.

The second method depends upon the fact that VCM absorbs infrared radiation at a number of wavelengths which are characteristic both of the unsaturation and of the carbon–chloride bond. The amount of radiation absorbed depends on the amount of VCM in the beam of radiation. In the equipment used in PVC plants, the optical system is designed so that the beam of radiation is repeatedly reflected between a pair of mirrors to give an absorption path length of more than 20 metres. This means that the equipment is of reasonable size, yet it can still detect concentrations of less than 1 ppm.

Both methods are in widespread use in PVC polymerisation plants, drawing in samples of air from a large number of points which are scanned sequentially every few minutes.

Such instruments are often used in conjunction with specially designed computers which permit rapid calculation of shift averages and analysis of excursions above hygiene standards. As an example of what is being achieved by such techniques, see paragraph 37 of the 'Vinyl Chloride Code of Practice for Health Precautions' issued by the Health and Safety Executive in February 1975.

3.3.2 Detection in Foodstuffs

VCM may occur in both PVC resins and articles fabricated from PVC where the concentration may range from fractions of 1 ppm to hundreds of ppm. However, in the case of foodstuffs the range is parts per billion. Quick, reliable methods of analysis were required to allow control of PVC manufacturing processes as well as to assess the progress in reducing the minute traces migrating into foodstuffs. The method which has gained general acceptance for such work is the so-called headspace method.[36] In this method, the sample is dissolved in a suitable solvent and allowed to equilibrate in a small vial. Gas from the headspace, consisting of a mixture of the solvent and VCM, is passed through a gas liquid chromatographic column and the peak due to VCM is measured. Systems are available which permit the automatic measurement of pre-loaded samples in a typical measurement time of thirty minutes. A number of solvents are used, such as tetrahydrofuran and dimethylacetamide.

In the case of PVC resins and objects, the method is capable of measuring fractions of 1 ppm of VCM. With some foodstuffs the method will detect a few parts per billion, whereas with others there are serious interference problems.

3.3.3 Portable Detectors

A number of miniature detectors have been developed which may be worn by plant personnel so that personal exposure levels can be measured directly. All types depend on the oxidation of VCM to hydrochloric acid or chlorine. These substances then stain specially treated paper, and the intensity of the stain gives an indication of the intensity of the concentration of VCM.

Such instruments will measure from fractions of 1 ppm up to 40 ppm, depending on the equipment.

A Chemical Industries Association publication[36] gives details of reliable analytical and detection methods.

3.4 CONCLUSIONS

Due to the long latency period of angiosarcoma of the liver, it is inevitable that some further cases will occur. However, there is no doubt that the major hazards associated with VCM have been corrected. Some further improvement in reducing atmospheric levels and the levels of VCM remaining in PVC resins will occur, but the improvements will be relatively marginal compared with what has already been achieved in so short a time.

PVC continues to remain an essential component of present day technology with an improved awareness of its remarkable versatility and usefulness.

ACKNOWLEDGEMENTS

The author gratefully acknowledges the help and encouragement of numerous colleagues in the PVC industry, both in the UK and overseas.

REFERENCES

1. VIOLA, P. L., BIGOTTI, A. and COPUTA, A. *Cancer Research*, **37** (1971), 516.
2. MALTONI, C., CRESPI, M. & BURCH, P. J. R. *Excerpta Medica International Congress*, Series No. 275, 1973.
3. BARNES, A. W. National and international aspects of the VCM health problem, Part II. Paper given at the *BPF/PRI Joint Conference: Vinyl Chloride and Safety at Work*, 28 May 1975.
4. *Proceedings of the Royal Society of Medicine Section of Occupational Medicine*, **69**, April 1976.
5. *Vinyl chloride toxicity and the use of PVC for packaging foodstuffs*. A presentation by the CEFIC Committee for the Toxicity of Vinyl Chloride. February 1976.
6. CRAMPTON, R. F. Assessment of toxic risks to workers in the plastics industry, *BPF/PRI Joint Conference Vinyl Chloride and Safety at Work*, 28 May 1975.

7. *VC/PVC: An Example of a Problem Resolved*, Verband Kunststofferzeugende Industrie e V Frankfurt am Main, West Germany, 15 September 1975.
8. Williams, D. M. J., *et al. Brit. J. Indus. Med.* **33,** (Aug. 1976) pp. 152–7.
9. Wilson, R. *Risk benefit analysis for toxic chemicals: vinyl chloride*. Paper given at the 172*nd ACS National Meeting*, San Francisco, 25 August 1976.
10. Schlatter, Ch. 'Gefährdung von Arbeitnehmer und Konsument durch Vinyl Chlorid.' Press Release by Lonza AG, 29 October 1975.
11. 'PVC and Health.' A background statement issued by the Society of Plastics Industry Inc., New York, April 1976.
12. British Patent 1 291 145.
13. British Patent 1 421 718.
14. British Patent 1 444 360.
15. Federal Register, 21 October 1976, **41,** No. 205. Environmental Protection Agency, National Emission Standards for Hazardous Air Pollutants, Standard for Vinyl Chloride.
16. Private communication.
17. Clayton, H. M. Guide lines for PVC processors. Paper given at the *BPF/PRI Joint Conference Vinyl Chloride and Safety at Work*, 28 May 1975.
18. Belgian Patent 833 558.
19. Belgian Patent 831 964.
20. Belgian Patent 833 041.
21. British Patent 1 407 665.
22. Belgian Patent 837 070.
23. British Patent 1 446 583.
24. US Patent 3 669 946.
25. Moffitt, T. W. Private communication.
26. Moffitt, T. W. Private communication.
27. Document 1565/VI/76-F.
28. Tester, D. A. and Moffitt, T. W. Vinyl chloride monomer in food packaging. Paper given at the *BPF/PRI Joint Conference Vinyl Chloride and Safety at Work*, 28 May 1975.
29. Gilbert, S. G. 'The migration of minor constituents from food packaging materials.' *J. Food Sci.*, **41,** 1976.
30. Abramowitz, R. J. 'Bulk polymerisation plants.' Paper given at the 172*nd ACS National Meeting*, San Francisco, 2 September 1976.
31. Unexamined Japanese Patent 47 076 (1976).
32. Unexamined Japanese Patent 47 081 (1976).
33. Unexamined Japanese Patent 17 950 (1976).
34. Unexamined Japanese Patent 17 287 (1976).
35. *Food and Cosmetics Toxicology*, **14,** No. 5, (1976) 498.
36. *The Determination of Vinyl Chloride*. A Plant Manual published by the Chemical Industries Association, Alembic House, 93 Albert Embankment, London SE1 7TU.
37. *VC/PVC: Health Protection Measures*. Verband Kunstofferzeugende Industrie e V Frankfurt. 12 November 1974.
38. *Colloque Scientifique sur le Chlorure de Vinyle Monomère*, tenu le mardi 2 Mars 1976, à l'Institut du Radium Fondation Curie, Paris.

Chapter 4

PVC ADDITIVES

W. V. Titow

Yarsley Research Laboratories Ltd, Ashtead, Surrey, UK

SUMMARY

In this chapter an attempt is made to indicate something of the nature and directions of the principal developments in the fields of the main PVC additives, viz. plasticisers, fillers and heat stabilisers. Reference is also made in a similar way to developments in what might be termed the minor additives, viz. polymeric modifiers, lubricants, flame and smoke retardants, colourants, UV stabilisers, foam aids and 'antistatic' additives.

The sheer size and diversity of the subject has necessitated a selective approach; however, within the scope available, mention has been made of many new technical developments and trends with a reference, in some cases, to associated economic factors, as well as reference to some recent work on the effects of additives and the mechanism of their action.

4.1 INTRODUCTION

The role of additives in PVC formulations is more important, and their functions more diversified, than in any other polymer composition. These are the factors largely responsible for the great variety and specialisation of materials used as PVC additives, and especially those major in importance and/or proportion incorporated, i.e. the plasticisers, heat stabilisers and fillers. The field is far too large for a comprehensive, in-depth review of the current state of its art and science (as well as the influence of commercial factors, highly significant in some sectors) to be attempted in a short paper. The treatment is therefore of necessity selective, and the presentation brief;

rather more emphasis has been placed on the technical than the commercial aspects of the developments and trends discussed.

4.2 PLASTICISERS

Recent developments and trends in this area may conveniently, if arbitrarily, be considered to fall into the following groups: new plasticiser types; extension of established ranges; and usage patterns.

4.2.1 New Types

4.2.1.1 *The Terephthalates*

Considerable interest was stimulated by the appearance in the USA of dioctyl terephthalate (DOTP), now available in the Eastman Kodaflex range, and comparable in price with many general-purpose plasticisers. This development stemmed from an evaluation by Eastman Chemical Products of a group of esters of terephthalic acid, with C_2 to C_{12} alcohols, as plasticisers for PVC. Among the principal results was the finding that, in general, a terephthalate ester behaves and performs like an orthophthalate ester of an alcohol with one more carbon atom in the alkyl chain.[1] The boiling point of DOTP is virtually identical with that of dioctyl phthalate DOP (383 °C and 384 °C respectively), but its volatility in PVC compounds is much lower. It also offers better nitrocellulose (NC) lacquer marring resistance, and gives greater viscosity stability in pastes. Some other performance properties are compared in Table 1. The fogging resistance of DOTP-plasticised formulations is good; it is said to pass the stringent new General Motors fogging test at 60 parts per hundred resin (phr).[2]

4.2.1.2 *Solid Ethylene Vinyl Acetate (EVA) Modified Polymers*

These materials are terpolymers, with ethylene and vinyl chloride as the two main components. Initially available from DuPont as development products ('Permanent Plasticiser Resins' PB 3041 and PB 3042) they are now marketed under the trade name 'Elvaloy'. These solid polymers, of reportedly high molecular weight (about 250 000), are used in proportions much higher than polymeric impact modifiers. They act as true plasticisers, albeit their efficiencies are comparatively low, and high-shear mixing is necessary for satisfactory compounding. Their most interesting performance characteristics in PVC compounds may be summarised as follows: their extractability (in detergent solutions and oils) and volatility are virtually nil as is their migration, and they have a high degree of

TABLE 1
PERFORMANCE OF DOTP AND DOP IN A PVC COMPOUND[a]

	DOTP		*DOP*		*DOTP*	
Plasticiser						
Parts per hundred resin	50		50		54	
Efficiency						
100% modulus, psi ($MN\,m^{-2}$)	1 600	(11·0)	1 500	(10·3)	1 500	(10·3)
Shore A durometer, 5 sec	83		80		80	
Mechanical properties						
Tensile strength, psi ($MN\,m^{-2}$)	2 900	(19·9)	2 875	(19·8)	2 850	(19·7)
Ultimate elongation, %	375		390		390	
Tear resistance, ppi	425		375		390	
Permanence properties						
Soapy water extraction, loss %	0·4		0·4		0·9	
Oil extraction, loss %	13		12		17	
Hexane extraction, loss %	26		26		28	
Volatility						
Activated carbon, loss %	0·8		1·4		0·9	
Retained elongation, 7 days at 100 °C, %	97		84		96	
Low temperature properties						
Torsion modulus, °C						
35 000 psi (241 $MN\,m^{-2}$)	−28		−26		−32	
135 000 psi (930 $MN\,m^{-2}$)	−60		−57		−67	
Brittle point, °C						
10 mil films (0·25 mm)	−35		−35		−38	
70 mil sheets (1·75 mm)	−29		−28		−39	

[a] Formulation (phr):
PVC resin 100
BaCd stabiliser 3
Plasticiser As indicated.

resistance to biodegradation. The heat-seal strength, tear resistance and low-temperature properties are also improved. The Elvaloys are compatible with most other plasticisers. Their main outlet will be in flexible and semi-rigid compounds. Wire coating, cable sheathing and upholstery are among the applications discussed.

4.2.1.3 *Other New Plasticiser Types*

A new type of polymeric (polyester) plasticiser, said to be chemically different from the conventional members of this general class, has recently been introduced by Ciba-Geigy into their 'Reoplex' range (Reoplex 1102).

Its most interesting performance features are its low viscosity and low thickening effect and viscosity stability in PVC pastes. These features are combined with low extraction and migration (characteristic of polymeric plasticisers) in the finished product. Some of these properties are illustrated in Table 2, in comparison with DOP, Reoplex 903 and Reoplex GL. Reoplex 903 is a medium molecular weight polyester plasticiser and Reoplex GL is a low molecular weight polyester plasticiser.

The ageing effect of some plasticisers in pastes has recently been related by Bigg and Hill,[3] on a quantitative basis, to the solvent power of the plasticiser for the polymer expressed in terms of an activity parameter α

$$\alpha = (1 - \chi)10^3/\mathrm{MW}$$

where χ is the Flory–Huggins interaction parameter and MW represents the molecular weight of the plasticiser. The correlation between the calculated values of α and viscosity increases (measured experimentally) is said to be very good.

A useful general review of the methods of assessment and expressions of the degree of interaction between plasticisers and PVC has been produced by Bigg.[4]

Recent polymeric plasticisers, based on urethane polymers, are typified by Ultramoll PU of the Mobay Chemical Corp., USA. This plasticiser is said to be compatible with PVC in any ratio and is also said to provide an extremely high degree of resistance to extraction and migration, combined with good low temperature strength.

4.2.2 Some Additions to Established Ranges

New polyadipate plasticisers (Palamoll 645, 647 and 855) have been added to the BASF range. Improved compatibility with PVC and monomeric plasticisers is claimed, together with improved stability to water extraction and outdoor exposure of the compounds.

Another new polyester product is Reoplex FG, introduced for use in formulations intended for food contact products (e.g. packaging film). Much lower extractability is claimed for this plasticiser in comparison with dioctyl adipate (DOA) in comparable film formulations.

Several materials in the Plastolein range (Emery Industries Inc., USA) exemplify recent additions or recently gained food-contact approvals (e.g. Plastolein 9775, 9789, 9051). The migration of some plasticisers (phthalate and epoxy types) from PVC compositions into edible oils was examined recently by a radioactive tracer technique.[5]

Successful formulation of rigisol pastes (plastisols with viscosity

TABLE 2

REOPLEX 1102—COMPARISON OF VISCOSITY EFFECTS IN A PLASTISOL FORMULATION

	Formulation			
	A	*B*	*C*	*D*
PVC paste polymer[a] (medium viscosity)	100	100	100	100
Reoplex 903	—	—	—	90
Reoplex 1102	—	—	90	—
Reoplex GL	—	90	—	—
DOP	90	—	—	—
Irgastab BC 206	2	2	2	2

	Viscosity (Brookfield) (poise)			
	A	*B*	*C*	*D*
At 23°C After 1 day				
5 rpm	50	156	272	648
10 rpm	41	159	280	716
20 rpm	33	167	298	828
50 rpm	24	191	344	>800
After 7 days				
5 rpm	50	152	280	632
10 rpm	45	152	272	672
20 rpm	35	157	290	760
50 rpm	27	174	328	>800

	Migration resistance (Refrigerator gaskets)	
	Reoplex 903	*DiDP*[b]
Weight loss (mg) of PVC disc at a plasticiser level of 100 phr.	4·1	267
Weight loss (mg) of PVC disc at a plasticiser level of 70 phr	3·7	171

[a] Viscosity number 130, determined by ISO method R174: 1961.

[b] DiDP = diisodecylphthalate.

sufficiently low for application, but gelling to a hard product) depends critically on plasticiser selection.[6] Whilst the principles of selection were formulated as early as 1960,[7] it is only comparatively recently that plasticiser producers have been publishing recommendations in this direction. An example is the Ciba-Geigy group of ester plasticisers for rigisols (Reomols DIDA, MN, MD and SD).

Another usage area for which additions have been made to existing plasticiser ranges is stain-resistant floor compositions. Examples from both sides of the Atlantic are Ciba-Geigy's Reomol SRB and Tenneco Chemicals' Nuoplaz 6186, 6187, 6188. A comparison of the resistance performance of a C_{79} phthalate against Reomol SRB is shown in Fig. 1.

An interesting attempt to computerise the formulation of plasticised compounds has been reported recently.[8] It involved the use of a programme covering a basic formulation of primary (phthalate) plasticiser 0–100 phr, extender (chlorinated paraffin) 0–100 phr and filler (calcium carbonate) 0–200 phr. Incompatible compositions are automatically excluded, and guidance provided on the cheapest formulations to meet a given set of requirements.

4.2.3 Usage Patterns and Changes

Since their introduction nearly a decade ago,[6] triaryl phosphates based on synthetic feedstocks have become firmly established and supply is expanding with demand. A Ciba-Geigy plant will shortly be going on stream at Trafford Park, Manchester, to manufacture these materials for the Ciba-Geigy Reofos range. The company is discontinuing altogether the production of the older tritolyl phosphate (TTP) and trixylyl phosphate (TXP) equivalents. These are still available in the Albright and Wilson Pliabrac range, although the supply of the synthetic-based Pliabrac phosphates is also expanding.

The use of 'straight-chain' 79 and 911 phthalates has gone on increasing under the continued influence of the combination of good service performance (including low-temperature properties, particularly good for phthalates) and favourable price factors. Their fogging effect in car upholstery is acceptably low in terms of most car manufacturers' specifications (e.g. General Motors, Ford, Volvo). New, more stringent tests are being introduced however; e.g. General Motors has increased the heating time at 71 °C from 6 to 72 hours. Although in tests of comparable or even greater severity the volatility performance of even the 79 phthalates can be better than that of, say, DOP or diisooctyl phthalate (DIOP)— and the 911 phthalates better than diisodecyl phthalate (DIDP) and didecyl

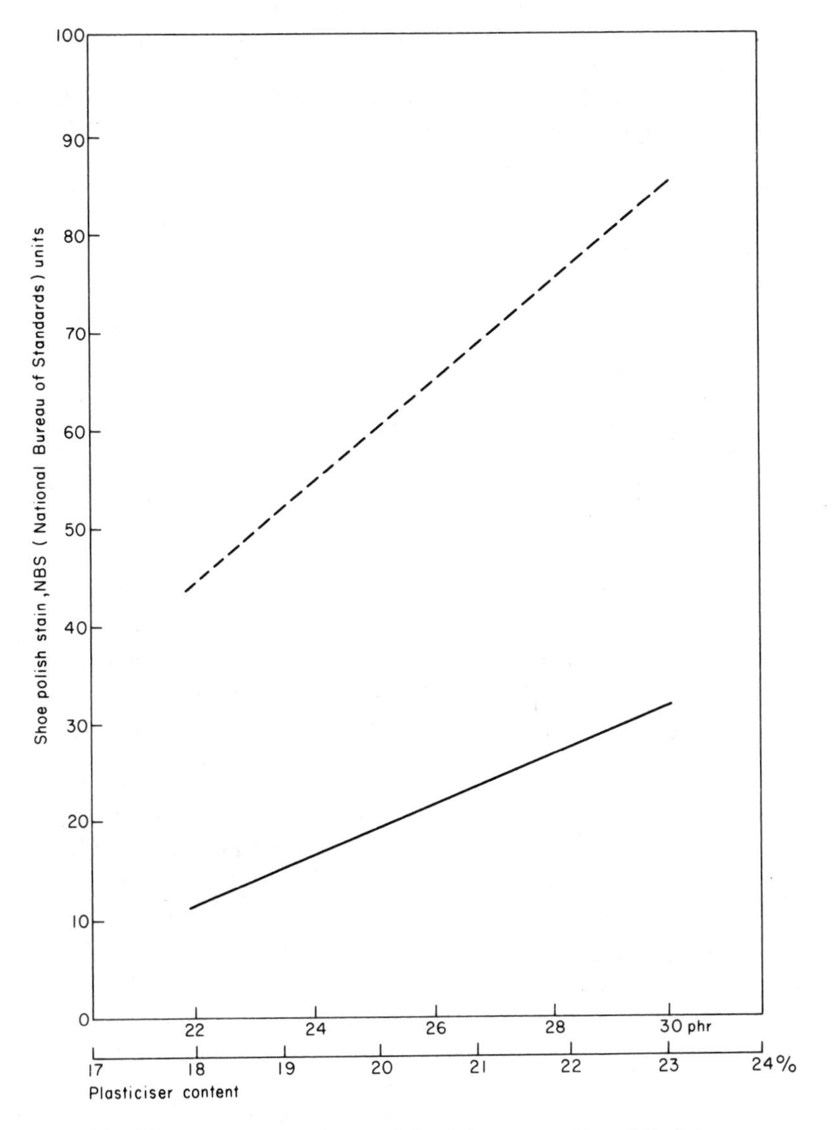

FIG. 1. Plasticiser content of shoe polish stain. ---- C_{79} phthalate; —— Reomol SRB.

phthalate (DDP),[6] the use is reportedly being considered of longer chain linear phthalates, trimellitates or even polymerics.[2] Increasing use of DIDP instead of DOP as a general purpose plasticiser is also reported from the USA.[2]

The properties of butyl benzyl phthalate (BBP) have always made it a popular candidate for use in compounds where good solvation, rapid fusion and lower processing temperatures are important. The high degree of interaction of this plasticiser with PVC, which is the factor responsible for these properties, makes it particularly suitable for use with resins processed to have low VCM contents. This is probably the main reason for a recent increase in demand for BBP.

4.3 FILLERS

Practical utilisation of the various effects of fillers and a widening understanding of the underlying mechanisms continue to demonstrate that the traditional distinction between reinforcing and cheapening fillers is one of functionality rather than kind. In PVC, as in other thermoplastics, improvements in some mechanical properties (e.g. stiffness, hardness, tear strength) can be produced by particulate fillers which are cheaper than the resin.[6] Glass fibres—the filler used for maximum practical reinforcement—are indeed considerably more expensive weight-for-weight, but only moderately so when filled compound cost is considered on the volume equivalent basis.[9] Some fillers used for special effects (e.g. electrical properties, machining characteristics, appearance) may additionally offer a degree of reinforcement or a cost advantage. The technical and commercial developments in the last few years have been concerned with several of the various filler functions.

4.3.1 Fibrous Fillers

4.3.1.1 *Glass Fibres*

Reinforcement with glass fibres significantly upgrades the mechanical properties of rigid PVC both short- and long-term (although the heat distortion remains low—see Table 3), and glass-fibre filled compounds have been available for several years.[9] A significant recent development in this area is a negative one—commercial production of the compounds is being curtailed or discontinued by some principal suppliers. This is mainly because the size of this market, never very large, does not warrant the

TABLE 3

EFFECT OF GLASS-FIBRE REINFORCEMENT ON SOME PROPERTIES OF RIGID PVC

Property	*Method of determination*	*Units*	*Unreinforced (Ethyl 7042 compound)*[a]	*Reinforced (Ethyl 7042 compound)*[a]		
				10% *glass*	20% *glass*	30% *glass*
Specific gravity	ASTM D-792		1·40 ± 0·02	1·45	1·53	1·61
Mould shrinkage	—	in. in^{-1} or mm. mm^{-1}	0·003–0·004	0·002	0·001	0·001
Tensile strength	ASTM D-638	psi ($MN\,m^{-2}$)	6 400 (44·1)	9 000 (62·0)	14 000 (96·5)	16 000 (110·3)
Tensile modulus	ASTM D-638	psi ($MN\,m^{-2}$)	420 000 (2 896)	650 000 (4 481)	1 200 000 (8 273)	1 300 000 (8 963)
Flexural modulus	ASTM D-790	psi ($MN\,m^{-2}$)	375 000 (2 585)	600 000 (4 136)	950 000 (6 550)	1 100 000 (7 584)
Impact-notch Izod $\frac{1}{8}$ in (3·175 mm) bar	ASTM D-256	ft lb in^{-1} ($J\,m^{-1}$)	15 (800·20)	6 (320·08)	3·5 (186·71)	2 (106·69)
Heat distortion at 264 psi (1·82 $MN\,m^{-2}$)	ASTM D-648	°C	70·6	81·7	86·7	87·8

[a] Ethyl Corporation, USA.

comparatively high cost of ensuring that the VCM content of the base polymer is kept within the low limits now mandatory, or becoming so, in industry.

Solely from the standpoint of reinforcement effects, carbon fibres would be more effective than glass fibres in PVC. However, their use in this polymer has never been an economically sensible proposition. In round figures, carbon fibre (chopped strand) costs about 60 times more than PVC resin; glass fibre costs about 1·7 times more. The cost of the base polymer in a compound containing, say, 25% carbon fibre is thus relatively unimportant. For applications calling for the highest performance (which would be the reason for using carbon-fibre reinforcement in the first place) a base polymer can be afforded with inherent 'engineering' properties better than those of PVC, e.g. nylon or polycarbonate.

4.3.1.2 *Asbestos*

The current preoccupation with the health-hazard aspect of the handling of asbestos in industrial processes, and its presence in products, extends to its role as a filler for plastics. Following the increased severity of the relevant OSHA (Occupational Safety and Health Administration) regulations in the USA, several American producers of phenolic moulding compounds are reported to be replacing asbestos by other fillers in their products.[2,10] The question has also been voiced[2] whether the established substantial outlet for short-fibre chrysotile (white) asbestos as a filler in vinyl flooring will be affected. In Britain this does not appear to be so.

The main suppliers, Turner Brothers Asbestos (TBA), are making their customers aware of the effects of the Health and Safety at Work Act 1974, but no fall in demand is reported. Neither are set-backs apparent in the second major outlet for chrysotile (in this application a long-fibre grade) in PVC, i.e. as reinforcement in pressed sheet material. TBA's own product (Duraform) has always been manufactured under strictly controlled conditions. There is no fall in demand either in connection with its main application (external and internal cladding, including flush-wall lining of food factories) or its various specialist applications (e.g. corrosion resistant trunking and ducting, and motorway signs). This situation reflects the fact that safe working with asbestos is possible if suitable precautions are observed.[11,12] It may also be noted that in its compounds with PVC the fibre is more intimately 'sealed' than in many non-plastics products, so that handling and fabricating (e.g. cutting) by the user is comparatively safer. In the case of Duraform, TBA Industrial Products and their official stockists also offer a cut-panel service.

No new major outlets for asbestos in PVC have arisen recently, although the use of chrysotile as filler in pipes has been under investigation. The main development activity is concentrated on improving still further the already good performance of asbestos-reinforced sheeting with regard to smoke and HCl generation by compounds: this is of particular interest in ducting systems for use with highly inflammable gases and liquids.

Another manifestation of concern for safety in handling is the recent appearance of Sylodex chrysotile asbestos fibres (W. R. Grace UK Ltd) in the form of 'crumb'. Whilst most of the chrysotile fibre fillers for plastics are of Canadian origin, this material comes from a particular deposit in California and is characterised by the flexible and highly micronised nature of the fibre. The 'crumb' for use with PVC is produced by wetting out with DIBP (diisobutyl phthalate 2 parts, to 1 of asbestos): this does not cause agglomeration or impair the effectivity of the fibre in its main application, i.e. as thixotropic additive to PVC pastes.

4.3.1.3 *Microfibre Fillers*

The same safety considerations which affect the use of asbestos may also influence the utilisation in PVC of Dawsonite. This is an hydrated sodium aluminium carbonate microfibre launched within the last few years by the Aluminium Company of America.[13] This filler has some reinforcing effect in rigid PVC (Table 4) where it also acts as a smoke suppressant and HCl-scavenger.[13] The main physical characteristics of Dawsonite are somewhat similar to those of Fybex, the potassium titanate whisker microfibre (Table 5), although Fybex is much heavier and, because of its higher refractive index, more effective as a pigment. Fybex-reinforced PVC compounds were available from several sources[9] until 1974 when Du Pont discontinued commercial production of this filler.

4.3.2 Particulate Fillers

Perhaps the most significant single trend in particulate filler technology in recent years has been the continued interest in surface treatment to improve dispersion in the resin and/or promote the desirable degree of adhesion at the filler/polymer interface.

Other points which may be noted are the inclusion of fillers in multi-additive packets (with stabilisers, lubricants and other components—cf. Section 4.4) for convenience of use in particular PVC formulations,

TABLE 4

PHYSICAL PROPERTIES OF DAWSONITE/RIGID PVC COMPOSITES

Property	*ASTM method*	*Units*	*Dawsonite level (% by weight)*			
			0	7·5	15	30
Melt flow rate	D1238-73F	g 10 min^{-1}	4·7	3·5	3·2	2·7
Deflection temperature	D648 at 264 psi (i.e. at 1·8 MN m^{-2})	°F (°C)	153 (68)	155 (69)	160 (71)	163 (73)
Coefficient of linear thermal expansion	D696	10^{-5} °C^{-1}	6·4	3·8	3·1	2·1
Izod impact strength (notched)	D256A	ft lb in^{-1} (J m^{-1})	3·39 (180·84)	3·00 (160·04)	2·62 (139·77)	2·36 (125·90)
Izod impact strength (unnotched)	D256E	ft lb in^{-1} (J m^{-1})	21·0 (1 120·29)	18·1 (965·58)	16·5 (880·23)	10·4 (554·81)
Tensile strength	D638	psi (MN m^{-2})	5 647 (38·9)	5 900 (40·6)	6 100 (42·1)	6 200 (42·7)
Tensile modulus	D638	10^6 psi (MN m^{-2})	0·41 (2 826)	0·54 (3 723)	1·43 (9 859)	1·60 (11 030)
Flexural strength	D790	psi (mN m^{-2})	11 760 (81·0)	12 420 (85·6)	12 900 (88·9)	13 160 (90·7)
Flexural modulus	D790	10^6 psi (MN m^{-2})	0·75 (5 170)	1·13 (7 791)	1·54 (10 620)	2·33 (16 060)

TABLE 5
SOME RELEVANT PROPERTIES OF DAWSONITE, FYBEX AND WHITE ASBESTOS (CHRYSOTILE) FIBRES

	Dawsonite	*Fybex*	*Chrysotile*
Natural fibre length (μm)	15–20	4–7	1 000–40 000
Fibre diameter (μm)	0·4–0·6	0·1–0·16	0·01–1
Density ($g\,cm^{-3}$)	2·44	3·2	2·55
Refractive index	1·53	2·35	1·50–1·55
Surface area[a] ($m^2\,g^{-1}$)	15–17	7–10	3–14

[a] BET N_2 method.

the use of wood flour in PVC extrusion compounds, and the developments in particulate flame retardants (see Section 4.5.3).

4.3.2.1 *Surface Treatments*

Progress continues in the stearate surface treatment of calcium carbonate fillers for PVC. Within the last few years most suppliers have introduced new grades representing developments in this area, for which better dispersion in PVC and improvements in compound properties are claimed.

Organo titanates have been receiving attention as coupling agents for fillers used in a number of thermoplastics, including PVC.[14,15] According to recent claims,[14] surface treatment of calcium carbonate filler with isopropyltri(dioctylphosphato) titanate or isopropyltri(dioctylpyrophosphato) titanate produces, *inter alia*, substantial improvements in impact resistance of 40% filled rigid PVC compounds and enables the amount of lubricant to be reduced. Beneficial use of the treated filler in PVC pipe compound has been suggested. In flexible PVC compositions the use of the first of these coupling agents is said to improve performance beyond that achievable with stearate-treated calcium carbonate as filler.[14]

It may be noted in passing that according to a recent report,[16] a large number of common PVC plasticisers and liquid additives were investigated as surface treatments for chrysotile asbestos fibre. A polyethylene glycol was said to have proved the most effective in promoting impact strength and flexural modulus in the filled PVC compound.

4.3.2.2 *Miscellaneous Fillers*

Extra fine or specially graded versions of calcium carbonate, claimed to give better processing and final compound properties in PVC, are also continuing to appear. The use of mica as a reinforcement filler for PVC has also been promoted recently (e.g. Suzorite, from Marietta Resources Ltd, in the USA).

Work directed to the development of new conducting PVC compositions has shown that the incorporation of 5–8 % by volume of nickel powder results not only in electrical conductivity increases but also significant increases in the strength of the filled material.[17]

Wood-filled PVC compositions for the production of extruded profile and sheeting are now available from more than one source.[18,19] They include 'Sonwood' from Sonesson Plast AB Sweden and 'Nordxyl' from Nordchem SpA Italy: a structural foam version of the 'Sonwood' product is also available. The filler does not produce any property improvements over those of the base polymer, but the specific gravity is reduced and the handling properties and performance are said to resemble those of wood. This is important in the main applications mentioned.

An ultrafine grade of micronised silica gel, highly effective as an anti-blocking agent in thin plasticised PVC film at 0·5 % addition level, has recently become available (Silica SM 111 from Joseph Crosfield & Sons Ltd).

4.3.3 Theory of Physical, Mechanical and Other Effects of Fillers in Filled Compositions

Interest in the theoretical aspects of the effects of fillers in filled polymeric composites continues. In an extensive review, Hale[20] has discussed the fundamentals of measurement and definition of the physical properties of composites and some of the ways of utilising existing theory and experimental data to predict such properties. Some theoretical and empirical models used in predictions of the thermal conductivity of filled polymer systems (including foams, regarded as polymer/gas composites) have been reviewed by Progelhof *et al.*[21] A discussion by Smith,[22] of the applicability of various theories to the prediction of the elastic constants of a composite from those of the polymer matrix and the filler, indicates that (for filler volume fractions up to 0·35) a modification of the van der Poel theory gives predictions in good agreement with experimental determinations. Smith's findings were extended by Dickie,[23] with special reference to the applicability of the theory to the prediction of the modulus of certain particle-filled composites. Nicolais[24] has produced an analysis of some mechanical properties of composites in terms of the relevant properties of their constituents. The effect of the filler packing fraction as a parameter influencing the properties of a filled plastic composition has been briefly considered by Ferrigno.[25]

4.4 STABILISERS

The concept of polyfunctionality, originally embodied to a limited degree in some coprecipitate stabilisers,[6] is now represented by other kinds of composite stabiliser products. Recent developments in pre-packed, multicomponent systems, which combine the stabiliser(s) with lubricants and fillers, are a significant extension of this concept. Such 'single pack' systems are designed for particular kinds of PVC compound, and formulated with due regard to the mutual compatibility of the constituents. They can offer cost savings (on stocks, storage and handling of the individual components) as well as operational convenience. They are of particular interest for small to medium scale in-plant compounding operations. The use of the single pack systems by a PVC compounder *ipso facto* involves some limitation upon the freedom and flexibility of formulation, but this may not be significant unless a large variety and volume of compounds are being produced and/or frequent formulation changes are involved.

The comparatively recent appearance of two new types of stabiliser, rivalling in some respects the established organotins, is another development of both technical and commercial interest.

4.4.1 New Types of Stabiliser

4.4.1.1 *The 'Estertins'*

The name 'Estertin' has been given by AKZO Chemie to a new class of organotin stabilisers, introduced within the last three years, and represented in the AKZO commercial range by four liquid stabilisers: Stanclere T208, T209, T215 and T217.

The estertins have the general structures:

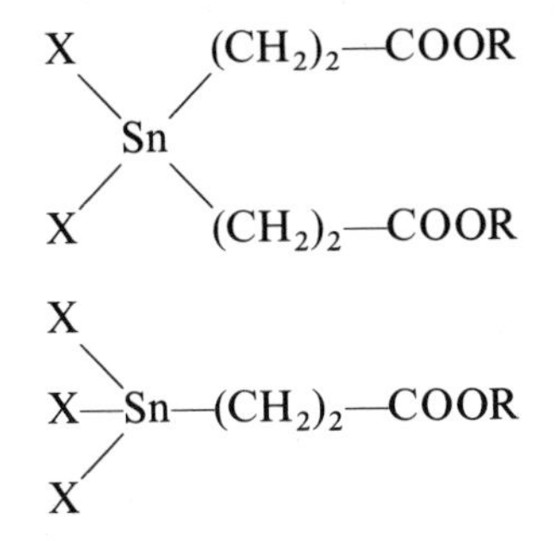

where X is a halogen (normally Cl) in the intermediates and a substituent group in the actual stabilisers.

In their production, use is made of tin chloride intermediates of low toxicity and volatility, formed by a new chemical route.[26]

Processing and service performance at least equivalent to that of existing alkyl tin stabilisers (with better light stability in some cases) are being claimed for the estertins in both pastes and compounds for, for example, pipe extrusion, rigid calendered sheet, extruded sheet and foil, injection moulding and bottles.[26]

4.4.1.2 *Antimony Mercaptides*

These materials, commercially available for about two years in the USA (originally from the Synthetic Products Co., now also from the Ferro Corp.) are liquid stabilisers consisting mainly of antimony trimercaptide.[27] Their handling characteristics are very similar to those of conventional liquid organotin stabilisers but their efficiency is greater at low content levels—up to about 0·8 phr, as shown in Fig. 2. So also is their stabilising

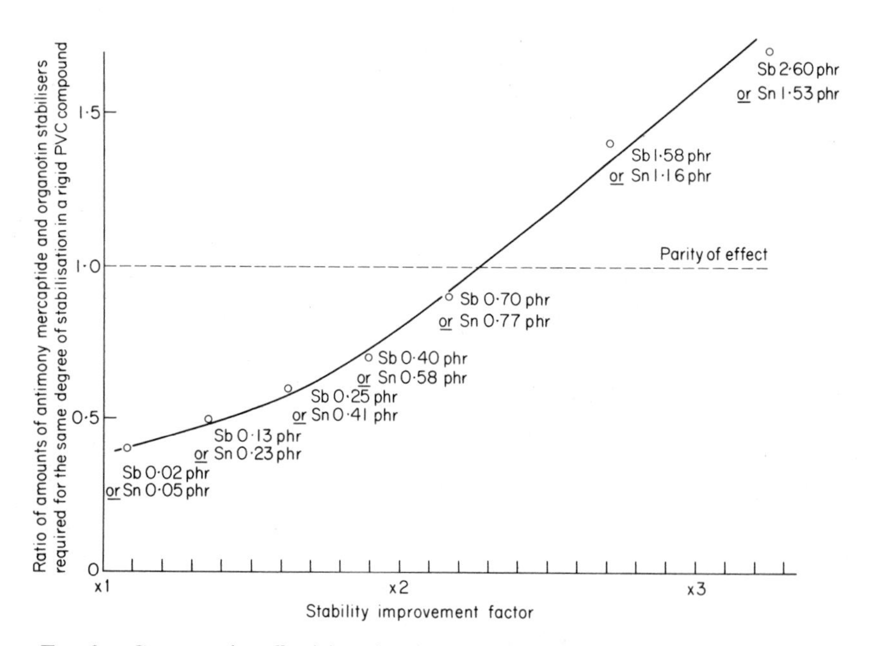

FIG. 2. Comparative effectivity of antimony mercaptide and organotin stabilisers in identical PVC compounds at generally low incorporation levels. Actual amounts of each stabiliser (to the first decimal place) required to give the equivalent stability improvements are shown by the individual points. Partly schematic representation based on suppliers' published data.

effect against heating at moderately high temperatures for long periods. The strong synergistic effect obtained with calcium stearate is noteworthy, as shown in Table 6.

Direct cost advantages in some formulations *vis-à-vis* the organotins are also claimed. However, the resistance to sulphide staining and to ultraviolet (UV) radiation (in clear compounds) is somewhat lower than that imparted by the organotins.

TABLE 6
CALCIUM STEARATE/ANTIMONY MERCAPTIDE SYNERGISM (MINUTES OF STATIC HEAT STABILITY)

Sb phr	*Ca phr*					
	0·0	0·1	0·25	0·5	1·0	2·0
0·0	0	0	0	0	0	0
0·1	10	10	15	20	30	35
0·25	20	20	20	30	40	50
0·5	30	35	40	50	60	70
1·0	50	60	70	70	95	95
2·0	90	110	110	125	140	95

The recent commercial appearance of antimony mercaptides as stabilisers for PVC is not an entirely new development, but an interesting revival promoted by improvements in PVC processing techniques and by the cost of materials. At the comparatively high use levels current in the early 1950s when their application in PVC was first explored, antimony mercaptides were less efficient than the organotins. Their greater efficiency at low levels is now coming into its own with the general lowering of stabiliser proportions used (in consequence of more effective processing on modern machinery, especially twin-screw extruders).

4.4.2 Other Stabiliser Developments and Trends

4.4.2.1 *Organotin Stabilisation*

Organotin stabilisers currently account for about 20% of the metal stabiliser market. Additions continue to appear in most manufacturers' ranges, for which performance or cost advantages are claimed. The cost advantage of methyl tins in comparison with butyl and octyl tins is being emphasised. Whilst the prices are closely similar, the methyl tins, richer in metal on a weight basis, can be used in lower amounts for comparable efficiency.

Interest in tin stabilisers suitable for food-contact uses also continues to be represented by new additions (e.g. Hoechst's dioctyl tin stabilisers VP Sn S40, 41 and 42; Cincinatti Milacron Chemicals' liquid organotin Advastab TM-181 FS).

Interesting suggestions have been made by Starnes and Plitz[28] for the pre-stabilisation of PVC. The PVC is reacted in a solvent with di(*n*-butyl)tin bis(*n*-dodecyl mercaptide), alone or in mixtures with di(*n*-butyl)tin dichloride, followed by recovery and purification of the polymer. Analysis showed that this purified polymer contained sulphur groups but only traces of residual tins; the resistance to thermal degradation could be increased almost ten-fold by this technique. It would appear that deactivation of vulnerable sites had been brought about by a reaction with mercaptide groups. The treatment may be of interest where stabilised, but substantially metal-free PVC is required.

4.4.2.2 *Lead Stabilisers*

Whilst lead compounds constitute the most important group among heat stabilisers, toxicity considerations have always been a factor in their handling and incorporation in PVC compounds.[6] Perhaps the most noteworthy developments are those combining improvements in these two areas with the continuing trend towards polyfunctionality in additives. Thus composite granular products are now available, equivalent in action to the most versatile coprecipitates, but containing standard materials. Multicomponent systems incorporating lead stabiliser(s), lubricant(s) and filler(s) have also reached the market pre-packed in polyethylene bags. They are suitable for direct introduction into a high speed mixer where the contents (with bag) are dispersed and compounded into the resin. The operational and hygienic advantages of this development are self-evident; the sealed bag method of compounding-in materials which may present a hazard on skin contact or when released into the atmosphere has been successfully used also with asbestos fillers.[9] Associated Lead Manufacturers, whose 'Almex' range includes products of both the granular composite and the single-pack, multicomponent type, are reported to be extending the second principle still further by incorporating a particulate flame retardant (antimony trioxide) into such systems.

4.4.2.3 *Alternatives to Cadmium-Containing Systems*

Among other metal stabilisers the recent additions to existing ranges reflect an interest in alternatives to cadmium-containing systems (especially cadmium/barium (Cd/Ba) and cadmium/barium/zinc (Cd/Ba/Zn)) for flexible compounds. This interest is apparently prompted by the comparatively high cost of cadmium compounds and an increasing preoccupation with the question of their degree of toxicity. For plastisols and calendering compounds, new barium/zinc (Ba/Zn) and calcium/zinc (Ca/Zn) liquid systems have been brought out (e.g. by Lankro Chemicals in Europe and Ferro Corp. in the USA) and strontium zinc (Sr/Zn) have also been introduced (e.g. by Ferro Corp. and Claremont Polychemical Corp. in the USA), as well as composite strontium/zinc/tin/phosphite systems (e.g. Nuostabe V-1925—Tenneco Chemicals, USA).

4.4.3 Stabilisation Mechanisms and Effects

Studies of interactions between PVC and stabilisers and their roles in the mechanisms of stabilisation, supplemented by work on thermal decomposition of PVC, continue to produce information relevant to a better understanding of this complex subject.

Considerable precision and facility of operation has been achieved through the use of a laser beam to decompose PVC specimens *in vacuo* directly in the source-cell of a mass spectrometer in a study of the course of decomposition and its products[29] (which were HCl:benzene:toluene in the ratio of 100:10:1). The presence of a plasticiser and antimony oxide was found to modify the decomposition significantly. Dehydrochlorination of PVC at 180–250 °C, also under vacuum, has provided data for what is claimed to be a simple mathematical model of the degradation process.[30] The thermal degradation of chlorinated paraffins used as secondary plasticisers in PVC was examined in comparison with that of the polymer.[31]

Some interesting correlations between the reactivity of stabilisers with HCl and their effectivity in PVC have been pointed out by Wypych.[32] The data cited link the high reactivity of organotin stabilisers with their high stabilising effects and indicate some possible reasons for synergism in other systems. The same author's inference[33] that epoxy compounds may act as agents transferring nascent HCl in decomposing PVC to the other heat stabilisers—which would act as the primary acceptors—appears to accord with one of Szabo and Lally's conclusions. From a world-wide study of the weathering of stabilised PVC[34] Szabo and Lally concluded that the presence of an epoxy plasticiser is beneficial to weathering resistance, but is more important with Cd/Ba systems than with tin stabilisers. These investigators also discuss the mechanism of action of stabilisers with special reference to their role in weathering.

Some general features of stabilisation relevant to the reprocessing of PVC scrap have been reviewed by Gleissner.[35] The relationship between thermal stability determinations with the aid of a Brabender Plasticorder and a capillary rheometer were discussed by Collins *et al.*[36]

4.5 OTHER ADDITIVES

4.5.1 Polymeric Modifiers

Polymeric modifiers have now been available for some time. Originally, such additives were used to improve the impact strength and the processability of PVC[6] but, thanks to continuing progress in this field, these effects have been combined with others so that improvements in heat distortion temperature and clarity have resulted. Heat distortion temperature may be improved by the use of acrylonitrile styrene copolymers alone or in conjunction with methacrylate butadiene styrene (MBS) systems. Some MBS systems impart particularly good clarity and this property is of course important in applications such as blown bottles, film and sheet.

Whilst no new major types of modifier have appeared in the last few years, polyalpha-methylstyrene (Amoco Resin 18) has been found useful as a processing aid and property modifier in PVC moulding and extrusion compounds.[37] Considerable improvements in toughness retention during weathering have been claimed[38] for a new impact-modified PVC resin (Geon 300 X6—B. F. Goodrich Chemical Co.). New proprietary products have also appeared within the existing chemical classes, including MBS polymers and acrylic processing aids (e.g. respectively, Paraloid KM 608 and T 25—Rohm and Haas). The effects and performance of some Blendex modifiers (Borg Warner Chemicals) were discussed fairly recently by Sahajpal.[39] A modifier reportedly based on a polyurethane polymer is being offered within the Landex range (Story Chemical Corp., USA).

An increase in the use of chlorinated polyethylene as a modifier (particularly in extruded expanded profiles) is attributable to the improvements it effects in rigid PVC's capacity to accept filler loading, and to the improvement in impact strength retention on weathering. The use of modifiers based on ethylene/vinyl acetate copolymers is also increasing, especially in Europe.

Since their first appearance in this form some years ago, the availability of nitrile elastomer powders has extended the scope and facility of production of PVC/nitrile rubber blends. The structure of blends of PVC with butadiene/acrylonitrile and its effect on glass transition in these materials were studied by Landi.[40] A computer-assisted study directed to the optimisation of blends of this kind for particular types of end use has been reported by Schwarz and Edwards.[41]

4.5.2 Lubricants

Whilst it is still fairly common to refer to internal and external lubricants, it has long been recognised[6] that the division is essentially a functional one, the type of function being basically governed by the compatibility of the lubricant with the resin. Thus, highly compatible materials will readily disperse in the PVC and provide internal lubrication by reducing intermolecular 'friction' (i.e. viscosity) in the molten state, whilst poorly compatible ones will tend to remain at the interface between the polymer and the processing surfaces, producing external lubrication.

Much attention has been given for many years to the lubrication requirements of various PVC compounds and to the effect on those requirements of stabilisers (some of which have lubricant properties[6]) and other components. The need for careful formulation, often involving the use of several lubricants, has been widely recognised.

In the last few years perhaps the most notable trend in this field has been the development and widening use of polyfunctional products, for both internal and external lubrication, including multicomponent lubricant systems, some of which combine anti-blocking and/or antistatic effects with the lubricating functions. Some of these systems form part of complete, single-pack additive formulations containing, as already mentioned, stabilisers and fillers. Much effort has been devoted by the suppliers to the development of these products which are aimed at particular areas of application—notably rigid extrusion compounds, especially pipe.

As with polyfunctional stabilisers, the main general advantage claimed for polyfunctional lubricants is convenience. The user is relieved of formulation problems and can use a single product catering for all the lubrication needs as well as some other requirements, resulting in product uniformity and storage space savings. Process cost and time savings, arising from a single addition to the PVC compound as against the need to separately weigh out and add perhaps several components, are further attendant advantages claimed. Some of the new polyfunctional lubricants also offer actual cost saving in the amount to be used.

Whilst the convenience and usefulness of polyfunctional lubricants in many directions are undoubtedly real, their use by definition deprives the compounder of the flexibility of making up his own polyfunctional lubricant formulations. The price of the proprietary products also includes the cost of the suppliers' blending, handling and packaging operations.

Apart from the polyfunctional products, individual additions to the suppliers' ranges have appeared in the last few years, especially among ester waxes and fatty acid derivatives, for which processing or cost advantages are claimed. Improvements in the purity, melting, flow, and dusting properties of calcium and zinc stearates are represented by new grades offered by some suppliers.

4.5.3 Flame and Smoke Retardants

Its high chlorine content makes the PVC resin (and hence also many rigid PVC compounds) self-extinguishing. However, compounds containing substantial amounts of combustible plasticisers will burn, producing voluminous smoke with a considerable HCl content.[42,43] Where phosphate and/or chlorinated paraffin extenders can be used, even plasticised compositions may have a sufficient degree of flame resistance. In other cases flame retardancy, and in particular smoke suppression, must be secured by incorporating retardants in the PVC compound.

Flame retardancy and smoke suppression are assuming increasing importance in PVC building materials, upholstery for car seating, cable and wire covering, wall coverings, carpet backing and many plastisol compounds in general. In the USA this is mainly because of the existence, and the continually increasing volume, of specific regulations for most of the above outlets. In Europe, where direct regulations and specific legal requirements are still much less numerous and prominent, consciousness of, and drive towards, greater fire safety is nevertheless present and taking practical effect. This has been fostered within the plastics industry itself and also 'externally' promoted by pressure from insurance companies.

Although no revolutionary new kinds of fire and smoke retardants have appeared recently, significant developments, mainly in retardant systems and their applications, have been taking place in the following areas.

4.5.3.1 *Antimony Trioxide*

This long-established, particulate flame retardant has recently been appearing in ultra-fine particle sizes (available from most suppliers) for which minimum opacifying effect and minimum influence on the desirable properties of PVC compounds are claimed.

The synergistic improvements in the use of antimony trioxide in conjunction with halogen/phosphorous compounds (e.g. dihaloalkyl phosphates) continue to be utilised. Other combinations are antimony/metallic silicate complexes (e.g. Oncor 75 RA—NL Industries in Europe, or Clarechem CLA-1500—Claremont Polychemical Corp. in the USA) and antimony trioxide with molybdenum trioxide. In this combination molybdenum trioxide is said to be a particularly effective smoke inhibitor (Fig. 3).

Another combination is antimony trioxide with zinc borate (up to equal parts). Although somewhat less effective as a flame retardant, the zinc

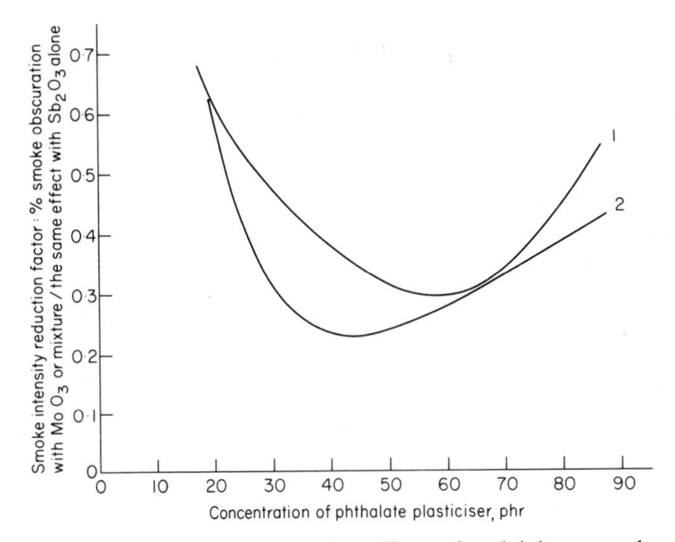

FIG. 3. Comparative smoke suppression effects of molybdenum and antimony oxides in the same PVC compound. *Curve* 1: Molybdenum v. antimony oxide at 6 phr concentration; *Curve* 2: Antimony/molybdenum oxide (1:1) at 6 phr total v. antimony oxide at 6 phr total. Partly schematic representation based on suppliers' published data.

borate is cheaper and also improves the light transparency of the compounds: the composition has been used in PVC wall coverings. The synergistic effect of zinc borate (not exhibited by barium and calcium borates) in combination with antimony trioxide has been confirmed in a study by Cowan and Manley,[44] who also found improvements in smoke suppression. These investigators point out, however, that the advantages of zinc borate as a flame retardant in PVC can be offset by the reduction in mechanical and electrical properties, as well as heat stability, unless small amounts of epoxidised oil are also included.

The finding regarding barium borate is at variance with recommendations that a commercial modified barium metaborate (Busan 11-M1—Buckman Laboratories, USA) is useful as a replacement for part (up to 50 %) of the antimony trioxide in PVC formulations. The advantages claimed are lower cost, lower opacity, improved weathering resistance and some fungicidal action in upholstery and outdoor fabrics. Addition of tetrabromophthalic anhydride (e.g. Firemaster PHT-4—Michigan Chemical Corp. USA) has also been suggested as beneficial for PVC compositions containing antimony trioxide.

4.5.3.2 *Alumina Trihydrate*

Considerable activity, especially as a smoke suppressant, has been claimed for this material in plasticised and rigid PVC formulations. Like antimony trioxide, alumina trihydrate is now available in very fine particulate form. Part of its smoke-suppressing action lies in its ability to absorb HCl fumes formed during the combustion of PVC.

4.5.3.3 *Other Metal Compounds*

Good results are claimed for the use of magnesium carbonate, in conjunction with phosphate ester plasticiser, as a smoke-suppressant in PVC wire coatings, carpet backing and wall coverings. One proprietary version of this system is the phosphate PHOSGARD LSV (Monsanto, USA). An inorganic magnesium/zinc complex (Ongard 1—NL Industries Inc. USA) is another example of a metallic system additive used for smoke suppression and flame retardancy in PVC.

4.5.3.4 *Fibrous Additives and Phosphate Plasticisers*

Reference has already been made (in Section 4.3.1) to the smoke-suppressing and HCl-scavenging activity of Dawsonite, the microfibrous, hydrated sodium aluminium carbonate. Asbestos fibres also exert a flame retardant effect if present in sufficient proportion in PVC compositions. Part of the already-mentioned expansion in the demand for phosphate plasticisers (Section 4.2.3) has been due to their flame retardant action. Their use for this effect, alone and in combination with chlorinated paraffin extenders, has been increasing: the presence of such extenders can considerably improve the low-temperature properties which are normally comparatively poor in phosphate-plasticised compounds.

4.5.3.5 *Flame-retardant, Low-smoke PVC Compounds*

A series of compounds in this category has been developed by B. F.

Goodrich. The first two to appear—Geon 87234 and 87235—combine mechanical properties normal for rigid PVC, with UL 94 V-O flammability ratings, high oxygen index, and smoke generation characteristics which satisfy the American FAA (Federal Aviation Administration) standards for aircraft cabins.

4.5.4 Colourants

In this field the accent is still very much on ease of handling and dispersion, important for both manipulation and cost economy. Trends are also continuing towards lower toxicity materials and those with reduced fire and pollution risks. Thus alternatives to the use of colourants based on cadmium and nickel are being explored and the utilisation of lead chromate pigments and molybdenum-based colours is also increasingly coming under scrutiny.

The various types of colour concentrate affording convenience of handling and ease of dispersion in PVC compositions are represented by the following:

1. *'Pre-blend' masterbatches in which the colourant is carried in solid polymer*. This is most commonly PVC or a vinyl chloride/vinyl acetate copolymer carrier (e.g. the Hoechst Hostavinyl pigment series). EVA is also sometimes used and cost savings over PVC-based concentrates can be effected with this carrier;
2. *Concentrate dispersions in plasticisers*. These are popular for use in plasticised PVC compositions. In some of the colour concentrates the carrier is a non-polymeric material, such as a fatty acid (e.g. the Ciba-Geigy Microlith range) or a wax lubricant (e.g. Hoechst Remafin).

In parallel with developments in the concentrates, improvements have been made in the particle shape, size and size distribution of pigments which represent the vast majority of colourants used in PVC. These improvements are again aimed at ease of processing, with particular reference to ready, complete dispersion.

Interest in fluorescent pigment preparations for PVC (which combine the already-mentioned ease of handling and dispersion with a non-dusting nature and better stability in service) is developing as it is for the same class of pigment in other polymers. The K500 series of pelleted fluorescent pigments (Hercules, Belgium) and the Sinloihi Color FR-50 series are examples of this kind of product.

4.5.5 UV Stabilisers

Oxidative photodegradation, caused by absorption of UV light in the presence of oxygen, is one of the principal mechanisms in the deterioration of polymers in weathering. The outdoor uses of PVC products (e.g. wall cladding, rainwater goods and window frames) are substantial, and growing: hence UV stabilisation is of particular importance in this area. Established UV absorbers, effective for the wave-band in the region of greatest sensitivity for PVC (about 310 nm), include 2(2′-hydroxy-5′-methyl-phenyl) benzotriazole (Tinuvin P, Ciba-Geigy) and some benzophenone compounds (e.g. Uvinul D49—GAF Corp.). Such compounds continue in demand.

A more recent introduction, of particular interest in transparent compounds and also compatible with optical whitening agents, is an oxalic anilide compound (Sanduvor VSU—Sandoz). Some sterically hindered phenolic compounds (e.g. Irganox 1076—Ciba-Geigy), which are essentially antioxidants, can augment the effect of primary UV absorbers (as well as heat stabilisers) in PVC compounds. A degree of UV stabilising effect in plasticised PVC has been claimed for zinc oxide.[2] Two UV absorbers based on cyanoacrylate are also available for use in rigid and flexible PVC formulations (Uvinal N-35 and N-559—GAF Corp., USA).

Attention has been drawn recently[45] to the role of what might be called the physical factors influencing the action of stabilisers, including UV stabilisers, viz. their degree of dispersion, migration through the polymer, extractability and volatility. Surface application of some UV absorbers to transparent PVC sheeting by absorption or in surface coatings was investigated[46] and found to be potentially cheaper, for comparable degree of protection, than the conventional incorporation into the polymer.

4.5.6 Foam Aids

Production of soft foams from plastisols, by mechanical gas entrainment (mechanical frothing), is a long-established process.[6] The importance of a suitable choice of surface active agent to promote frothing and stabilise the foam is illustrated by the results of a recent study[47] and demonstrated by the results obtainable with the aid of silicone surfactants specially developed (by the Dow Corning Co.) for this application.[48]

Other new proprietary products in the PVC foam field include (i) cell control agents (some particularly effective in heavily filled compounds—e.g. VS-103 Air Products and Chemicals Inc., USA); (ii) chemical blowing agents with reduced plate-out effects (e.g. modified azodicarbonamides

Kempore FF and MC—Stepan Chemical Co., USA), and (iii) stabiliser/activators for chemically blown foams.

4.5.7 Antistatic Agents

Proprietary antistatic agents of comparatively recent vintage include Irgastat 51 (Ciba-Geigy) for plasticised PVC, and Sandvin VU (Sandoz), a solid (powder) compound, acceptable for food-contact in packaging materials, stable at PVC processing temperatures, and with long-lasting action. The results of a study, by Sheverdyaev *et al.*,[49] of thermo-oxidative degradation of PVC containing antistatic agents, suggest that the dehydrochlorination and oxidation of the polymer is intensified if the additive contains nitrogen.

ACKNOWLEDGEMENTS

For a few individual items of information for Sections 4.2–4.4 the author is indebted to Mr B. J. Lanham of LNP Plastics Nederland BV, Mr M. Combey of Ciba-Geigy, Mr R. R. Bowman of Joseph Crosfield & Sons Ltd, Mr J. C. Cornforth of W. R. Grace Ltd, Mr L. F. Parks of Cape Asbestos Fibres Ltd, Mr R. F. Sheppard of TBA Industrial Products Ltd, and Dr P. S. Coffin of Associated Lead Manufacturers Ltd.

Tables 1, 4 and 6 are reproduced with permission of the SPE from references 1, 13 and 27 respectively. Table 2 and Fig. 1 are reproduced by Courtesy of Ciba-Geigy (UK) Ltd from their technical literature.

REFERENCES

1. Bealer, A. D. *SPE 34th ANTEC Proceedings*, 1976, 613.
2. Naitove, M. H. and Evans, L. *Plastics Technology*, 1976, **22**(8), 49.
3. Bigg, D. C. H. and Hill, R. J. *J. Appl. Polym. Sci.*, 1976, **20**(2), 565.
4. Bigg, D. C. H. *J. Appl. Polym. Sci.*, 1975, **19**(11), 3119.
5. Kampouris, E. M. *Polym. Engng. Sci.*, 1976, **16**(1), 59.
6. Penn, W. S., *PVC Technology*, 3rd ed., Titow, W. V. and Lanham, B. J. Eds., 1971, Applied Science Publishers Ltd, London.
7. Goodier, K. *Proceedings of the International Congress on the Technology of Plastics Processing*, 1960, NV't Raedthuys, Amsterdam.
8. Pugh, D. M. and Wilson, A. S. *Eur. Plast. News*, 1976, **3**(9), 37.
9. Titow, W. V. and Lanham, B. J., *Reinforced Thermoplastics*, 1975, Applied Science Publishers Ltd, London.
10. Anon., *Resin News*, 1976, **16**, 2.

11. *Asbestos Dust—Safety and Control*, Asbestos Information Committee, London 1976.
12. *Control and Safety Guides*, Nos. 1–9, The Asbestos Research Council.
13. BONSIGNORE, P. V. *SPE 34th ANTEC Proceedings*, 1976, 472.
14. MONTE, S. J., SUGERMAN, G. and SEEMAN, D. J. *Ibid.*, 35.
15. MONTE, S. J. and SUGERMAN, G. *Ibid.*, 27.
16. AXELSON, J. W. and KIETZMAN, J. H. *Ibid.*, 601.
17. KUSY, R. P. and TURNER, D. T. *SPE Journal*, 1973, **29,** 56.
18. ANON., *Mod. Plast. Internat.*, 1976, **6**(10), 12.
19. ANON., *Europ. Plast. News*, 1976, **3**(10), 49.
20. HALE, D. K. *J. Mat. Sci.*, 1976, **11,** 2105.
21. PROGELHOF, R. C., THRONE, J. L. and RUETSCH, R. R. *Polym. Engng. Sci.*, 1976, **16**(9), 615.
22. SMITH, J. C. *Polym. Engng. Sci.*, 1976, **16**(6), 394.
23. DICKIE, R. A. *J. Polym. Sci., Polym. Phys. Edn.*, 1976, **14**(11), 2073.
24. NICOLAIS, L. *Polym. Engng. Sci.*, 1975, **15**(3), 137.
25. FERRIGNO, T. H. *SPE 34th ANTEC Proceedings*, 1976, 606.
26. LANIGAN, D. *Krauss-Maffei 5th International Extrusion Symposium*, Linz/Asten, 8–11th September 1976.
27. DIECKMANN, D. *SPE 34th ANTEC Proceedings*, 1976, 507.
28. STARNES, W. H. and PLITZ, I. M. *Macromolecules*, 1976, **9**(4), 633.
29. LUM, R. M. *J. Appl. Polym. Sci.*, 1976, **20,** 1635.
30. TROITSKII, B. B., SOZOROV, V. A., MINCHUK, F. F. and TROITSKAYA, L. S. *Eur. Polym. J.*, 1975, **11**(3), 277.
31. SOSA, J. M. *J. Polym. Sci., Polym. Chemistry Series*, 1975, **13**(10), 2397.
32. WYPYCH, J. *J. Appl. Polym. Sci.*, 1976, **20**(2), 557.
33. WYPYCH, J. *J. Appl. Polym. Sci.*, 1975, **19**(12), 3387.
34. SZABO, E. and LALLY, R. E. *Polym. Engng. Sci.*, 1975, **15**(4), 277.
35. GLEISSNER, A. *Kunststoffe-Plastics*, 1975, **22**(1), 11.
36. COLLINS, E. A., METZGER, A. P. and FURGASON, R. R. *Polym. Engng. Sci.*, 1976, **16**(4), 240.
37. WILSON, A. P. and RAIMONDI, V. V. *SPE 34th ANTEC Proceedings*, 1976, 513.
38. SUMMERS, J. W. *Ibid.*, 333.
39. SAHAJPAL, V. K. *Kunststoffe*, 1976, **66**(1), 18.
40. LANDI, V. R. *Acrylonitrile in Macromolecules, Applied Polymer Symposium No.* 25, 1974, 223, John Wiley.
41. SCHWARZ, H. F. and EDWARDS, W. S. *Ibid.*, 243.
42. CALCRAFT, A. M., GREEN, R. J. S. and MCROBERTS, T. S. *Plastics and Polymers*, October 1974, 200.
43. WOOLLEY, W. D., *Plastics and Polymers*, December 1973, 280.
44. COWAN, J. and MANLEY, R. T. *Brit. Polym. J.*, 1976, **8**(2), 44.
45. ALLARA, D. L. *SPE 34th ANTEC Proceedings*, 1976, 245.
46. KATZ, M., SHKOLNIK, S. and RON, I. *Ibid.*, 511.
47. DEANIN, R. D., CALCUTTAWALA, A. M. and CAPASI, V. C. *Ibid.*, 501.
48. ACTON, J. and DEBAL, F. *Plastics & Rubber Weekly*, 18 June 1976, 22.
49. SHEVERDYAEV, O. N., SLESAREV, V. V., TUGOV, I. I., PUDOV, V. S. and ZHURAVLEV, V. S. *Plast. Massy*, 1976, **6,** 53.

Chapter 5

THE CONTINUOUS COMPOUNDING OF PVC

P. RICE and H. ADAM
Werner & Pfleiderer (UK) Ltd, Stockport, UK

SUMMARY

This chapter points to the difference in melt behaviour and consequently to the processing characteristics of PVC compared with other thermoplastics. Reference is made to: effect of heat and shear; method of pre-mixing; pros and cons of using dry-blend (powder) and compound (pellets); choice of compounding and pelletising machine systems; characteristics of twin screw design; details of a complete compounding plant; economy of in-plant compounding. The importance of the compounding operation is emphasised as it directly influences both material economics and end-product properties.

5.1 INTRODUCTION

Polyvinylchloride (PVC) was one of the first mass-produced synthetic plastics and still continues to find new markets. Partly because of the well-known processing difficulties of the basic resin it is rarely used in its pure form. Fortunately, the resin lends itself very easily to modifications such as the addition of plasticisers, heat stabilisers, fillers or polymeric modifiers. All PVC materials must therefore undergo compounding in order to achieve good material and processing properties. Ideally, such compounding should impart the minimum heat history, and the cost must be reasonable.

5.2 MELT BEHAVIOUR OF PVC

There are three major reasons why PVC in its pure form presents serious problems to the processor: we will deal with them in turn.

5.2.1 Mechanical Properties

Unmodified PVC in melt form is a pseudo-plastic material and exhibits very non-Newtonian characteristics, as its apparent viscosity is very much dependent on shear rate (Fig. 1). The power law

$$\tau_s = K\left(\frac{d\gamma_s}{dt}\right)^n$$

where τ_s = shear stress, K = a constant, $d\gamma_s/dt$ = shear rate, and n = flow index, approximately describes these curves. The exponent n is an indication of flow properties and its value ranges between 0·4 and 0·6 for

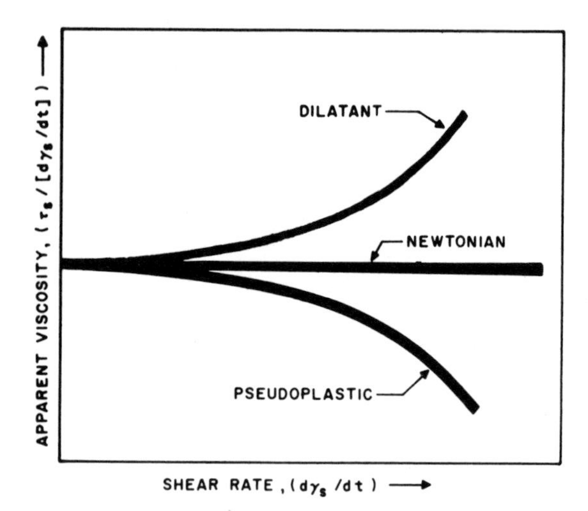

FIG. 1. Apparent viscosity (shear stress/shear rate) v. shear rate of different fluids.

most polymer melts. PVC with a flow index of 0·3 would flow through a tube in the way shown in Fig. 2. In other words, as the exponent n decreases in value, we are approaching plug flow conditions with all or most of the shear taking place in a very thin layer next to the confining walls. There is little or no shear within the mass of the material, i.e. in the centre of the flow channel.

For PVC to be used in its unmodified form, the processing equipment and the conditions of use must overcome the problem of high shear rate near the confining walls (barrel or screw surfaces).

This is one reason why PVC in its unmodified form is so difficult to process. When the polymer is introduced into a processing apparatus and starts to soften, it becomes very difficult to control the shear conditions. There is the tendency for extremely high shear conditions, with subsequent overheating of the material, to develop immediately near the confining walls of the extruder while at the middle of the extruder channel there is practically no shear. When PVC was first processed the flow index was

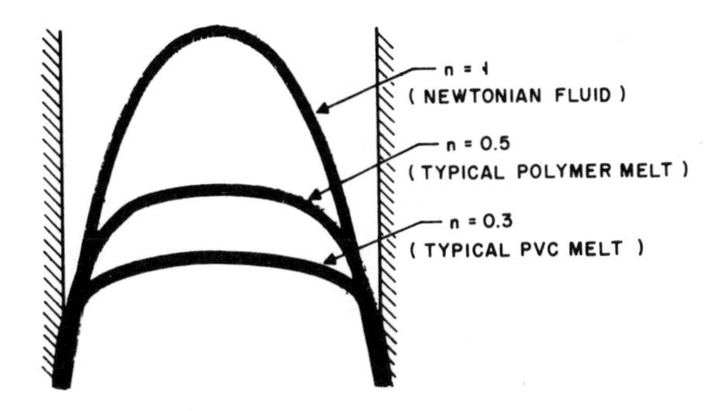

FIG. 2. Typical velocity profiles for melts within a channel or tube.

brought from 0·3 to 0·5, a value typical for most polymers, by using plasticisers; polymeric modifiers may now also be used. The use of substantial quantities of plasticisers, however, has a very marked effect on final material properties, especially stiffness and tensile strength, and the use of polymeric modifiers can also create similar undesirable effects (apart from the fact that they represent an additional substantial cost element).

5.2.2 Effect of Temperature

PVC resin as such is largely amorphous, and it is difficult to establish its precise melting point. Its properties on heating are not very well defined compared with those of crystalline polymers, and such studies are complicated by the thermal instability of PVC. The material slowly softens as the temperature rises (see Fig. 3), but before the melt stage is reached, the material degrades. Even with the best stabilisers, PVC cannot practically be processed at temperatures above 200 °C and this means that PVC is processed at temperatures which are significantly below the melt stage. The low glass transition temperature of PVC, 75 °C, should also be noted.

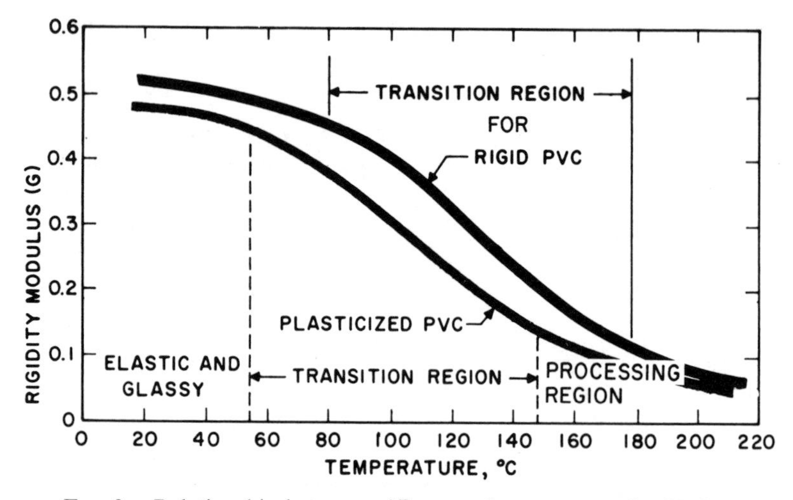

FIG. 3. Relationship between stiffness and temperature for PVC.

5.2.3 Thermal Instability

This brings us to the third major element in PVC processing, which is the tendency of this material to degrade by an auto-catalytic reaction. The degradation activation energy (approximately 20 kcal/mol for PVC) is substantially less than the specific energy required to generate a melt. In other words, it is difficult to arrest initial degradation reactions when PVC is heated. The seriousness of this problem can be illustrated by comparing the degradation activation energies of PVC with other major polymers: polystyrene—55 kcal/mol; polyethylene—46 kcal/mol; polypropylene—65 kcal/mol. These major PVC melt phenomena, by necessity, dictate the design and operation of commercial conversion equipment. For example, to overcome the problems introduced by plug flow, i.e. with all or most of the shear taking place in the very thin layer next to the boundary wall, it is necessary to use thin material layers. If this is not done, the vast bulk of the polymer in the flow channel is not subject to shear and poor mixing will therefore result. For the same reason, mechanical energy input cannot be solely relied upon for melting, as any heat generated is confined to the thin layer wherein the shear is concentrated. This means that we cannot operate PVC conversion equipment adiabatically.

Now, due to the melt characteristics of PVC, we must operate the process substantially below the temperature at which the polymer becomes a melt and at the same time we must provide sufficient shear throughout the mass of the material so as to properly disperse the polymer additives. However, careful consideration must be given to the stock temperature achieved so as to avoid degradation and keep stabiliser concentrations at economically reasonable levels.

The compounding stage is extremely important, and the compounding conditions must be chosen to suit a particular formulation. If PVC is compounded insufficiently, the final extrudate will show surface defects and poor dimensional stability. These faults can be traced directly to incomplete heating of the polymer and inadequate dispersion of filler, modifiers and other additives. If too low a melt temperature is chosen, the PVC particles will not completely fuse and completely uncompounded areas will therefore exist in the final product. If the temperatures chosen are too high, the chances are that the melt viscosity will be brought down to a point where insufficient internal shear is generated in commercial equipment. There will be relatively thick material layers and the ever-present danger of polymer degradation will arise. While most of these observations refer to rigid PVC formulations, similar considerations prevail for plasticised and highly filled formulations: these can greatly affect, for example, the line speeds of cable coating operations.

5.3 POWDER v. PELLETS

When considering setting up an extrusion operation for PVC, the first thing that must be decided is whether to use direct powder processing (made by pre-mixing), or pellets.

There are two types of pre-mixer; slow and fast. Ribbon blenders, with a jacketed, steam heated wall are typical of slow mixers. These take large batches, say 300–500 kg, and have a mixing/heating time of 30–45 minutes. Due to their large volume, they need few weighing operations but require a large holding tank between the mixer and compounder.

Fast mixers are turbo mixers with typical batch weight of 100–150 kg and mixing/heating time of 5–8 minutes. The heat is generated mechanically by the fast moving rotor blade. Fast mixers are easy to clean but require a much more powerful motor. In recent years, this type of mixer has become the predominant choice for most PVC compounding operations. Direct powder processing has advantages in some types of processing, for instance, high volume production of pipe.

Direct powder processing offers the following advantages:

1. Material preparation is more economical since the pre-mix does not have to be converted to a gelled compound;

2. The heat history of the material is reduced, as it is heated only once; and
3. Only the basic raw materials need be stored and the desired formulations can be made at any time by a relatively simple blending operation.

However, if pre-mixes are used product quality can suffer because

1. Optimum dispersion of ingredients and complete gelling are sometimes not achieved;
2. Devolatilisation of the material is difficult; and
3. Temperatures are difficult to control.

It must be emphasised that the key to a successful powder operation is consistency of the pre-blends.

In PVC applications, where product standards call for improved compound quality and uniformity, pelletising has become almost mandatory. To the processor with more than one product line, it offers the advantage of flexibility, as efficient utilisation of a compounding line is achieved.

The advantages of pellets are that they:

1. Contain all components and additives homogeneously dispersed—as a result conversion operations can be run at higher line speeds;
2. Have excellent feeding characteristics because of their high and constant bulk density;
3. Are free of moisture and other volatiles—these could cause voids in the end product;
4. Are free of dust and sticky particles—minimum cleaning is therefore required during colour changes; and
5. Have excellent storage and conveying characteristics.

5.3.1 Reduction of Vinyl Chloride Monomer (VCM)

Very significant improvements have been made in recent years by the polymer manufacturers in reducing the level of residual VCM in the powder supplied for compounding or direct conversion into profiles etc. Typical VCM content in the polymer nowadays is as low as 20–30 ppm. If it is desired to reduce these levels even further, then proprietary equipment is available for use with high-speed mixers. The mixer is purged with fresh air during the time the polymer powder is being heated and fluidised (i.e. in the best state to expose the maximum surface). The VCM concentration can be reduced to less than 5 ppm by such techniques.

5.4 MACHINE DESIGN CONSIDERATIONS

When selecting new compounding equipment, it is now possible to choose from a number of design solutions which provide means for controlled shear input and large heat transfer surfaces.

5.4.1 Single-screw Systems

The following systems are available:

1. Single-screw, single-stage, non-vented extruder;
2. Single-screw, two-stage, vented extruder; and
3. Reciprocating-type mixing extruder.

The best-known example of the last is the Ko-Kneader, but all are used successfully for plasticised and filled PVC formulations. With these types of compounders, shear can be introduced into the polymer melt over a fairly long section of the compounder barrel, thus satisfying one of the basic conditions of PVC compounding, i.e. spreading the energy-input areas.

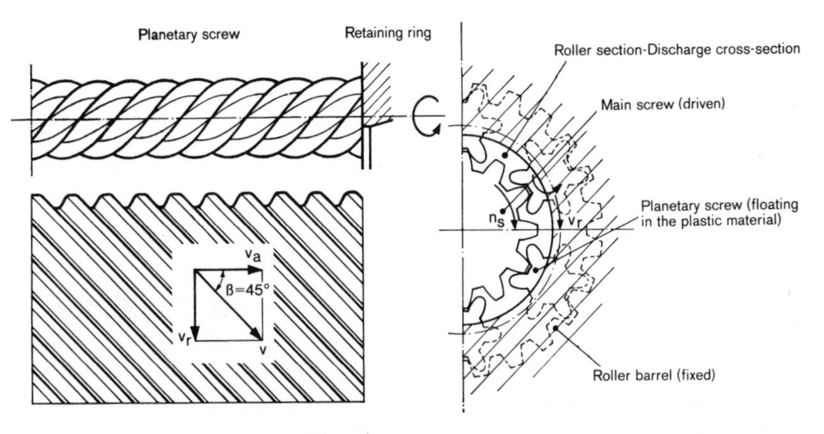

FIG. 4. The planetary screw system.

Good use has also been made of the so-called planetary screw system (Fig. 4) which, strictly speaking, is not a single-screw system, whereby shear is introduced, for melting and compounding, by relatively small diameter intermeshing screws rotating around the main screw. With this system, a limited degree of self-cleaning can be achieved which in turn helps to prevent degradation.

5.4.2 Twin-screw Extruders

Twin-screw extruders have definite advantages over single-screw machines and, in the past few years, have become widely accepted for PVC compounding. The twin-screw extruder allows the processor to maintain extremely close control over all process variables including stock temperature, shear, residence-time distribution and product uniformity (Fig. 5). In addition, multiple-port feeding and single or multi-port devolatilisation are easily accomplished with twin-screw machines.

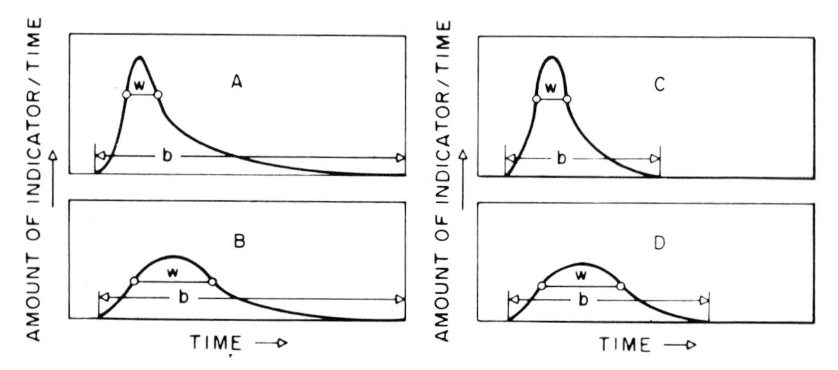

FIG. 5. Four typical residence–time curves for twin-screw extruders. w: Degree of the longitudinal mixing effect of the extruder screw; b: Degree of self-cleaning effect of the extruder screw; A: Extruder with poor longitudinal mixing and poor self cleaning; B: Extruder with good longitudinal mixing and poor self cleaning; C: Extruder with poor longitudinal mixing and good self cleaning; and D: Extruder with good longitudinal mixing and good self cleaning.

By equipping twin-screw machines with screw kneading elements, it has become possible to introduce very high amounts of energy at predetermined locations in the material flow path. These make it possible to mix components with very different viscosities, to closely control shear (while reaching shear rates up to $10\,000\ s^{-1}$), and to change the direction of shear during processing.

Another advantage of twin-screw compounding extruders is that they can be operated with the screw flights only partially filled with material without causing surging. The ability to starve-feed these machines eases the task of devolatilising, where the product must pass beneath an open port at zero pressure.

Because twin-screw extruders impart less shear and compression to rigid PVC compounds, less stabiliser and lubricants are required and this

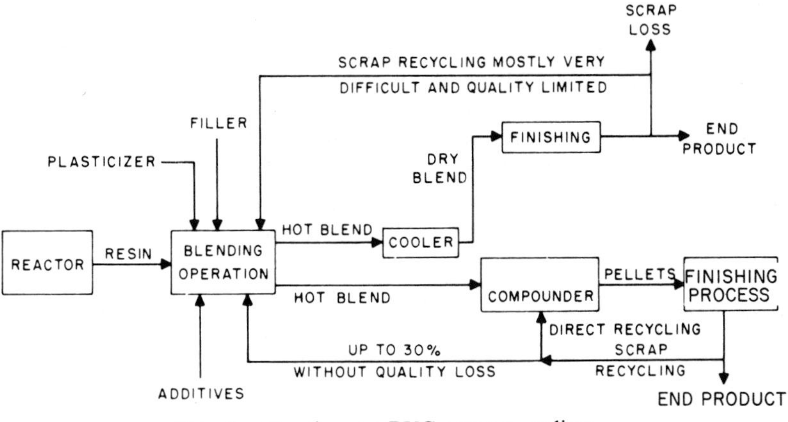

FIG. 6. Continuous PVC scrap recycling.

significantly reduces the cost of the overall formulation. For example, in comparing pipe grade compounds, the cost differential between formulations required for single and twin-screw machines can be as much as 30%.

Another economic advantage of twin-screw compounding systems is the ability to easily recycle scrap without quality loss (Fig. 6). Because of the uniform heat history to which each particle is subjected, and because the processing can be accurately controlled, up to 30% scrap can be re-compounded without affecting output rates. Figure 7 shows the effect of using 100% regrind material.

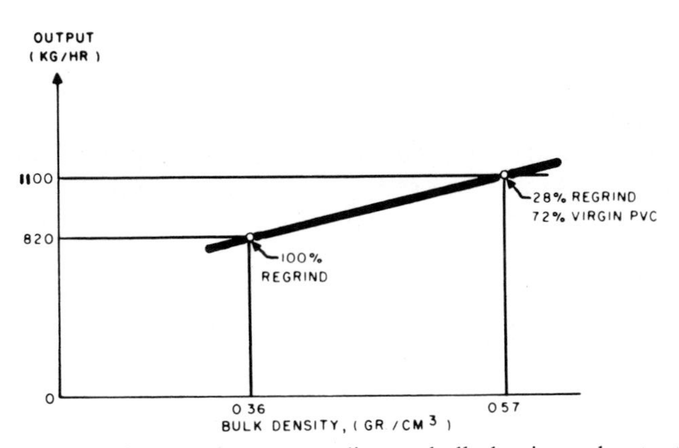

FIG. 7. Influence of scrap recycling on bulk density and output.

5.4.3 Two-stage Compounding Systems

Owing to the particular problems in compounding PVC, and other heat and shear sensitive materials, recent developments have resulted in compounders which combine the mixing and process control advantages of twin-screw extruders with the conveying advantages of single-screw machines. This is accomplished by separating the mixing and compounding operation from the pumping, conveying and pelletising function (Fig. 8). The new two-stage systems are equally effective with both rigid and flexible PVC formulations.

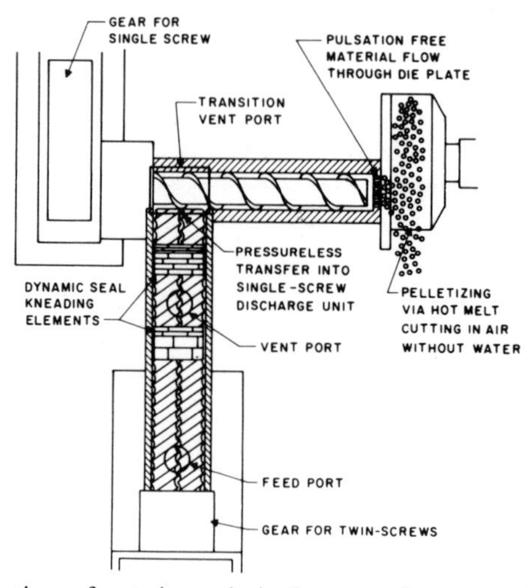

FIG. 8. Combination of a twin and single screw for processing heat sensitive material.

Material degradation is avoided by heating the material in a short (L/D—15:1) twin-screw extruder (co-rotating at 300 rpm) under relatively high shear. It is then transported into the pressureless area of a slowly rotating (21 rpm) single-screw with a 6:1 L/D ratio. In operation, preblend is introduced into the feed section of the twin-screw compounder where high pitch screws maximise the intake and pump the feed, in a very positive manner, through three separate barrel zones where the compound temperature is raised by externally supplied heat (Fig. 9). The heated PVC then passes through two kneading zones.

By using interchangeable screw and kneading elements, the length and width of these sections can be varied so as to maintain the correct heat levels, necessary for dispersion of additives, without degradation of either the

TABLE 1

TYPICAL OPERATING CONDITIONS OF A TWO-STAGE COMPOUNDING SYSTEM

Main drive rating, kW		*Average hourly output of PVC*			
Twin-screw section	*Single-screw section*	*Plasticised*		*Rigid*	
		kg	*lb*	*kg*	*lb*
15	10	280	616	165	363
42	27	800	1 760	480	1 056
80	31	1 500	3 300	900	1 980
160	48	2 800	6 160	1 800	3 960
240	72	4 400	9 680	2 500	5 500

additives, such as blowing agents, or the polymer. The material then passes into the single-screw extruder where it is conveyed through a breaker plate for die face pelletising. Vacuum can be applied to the transition chamber for removal of volatiles.

The single-screw section of the compounder need not raise the temperature of the material, and it can be equipped with barrel and screw cooling to lower the temperature if required. Since the temperature can be accurately controlled, the two-stage machine is suited to automatic, in-line, air-cooled pelletising, which is recommended for the production of moisture-free pellets.

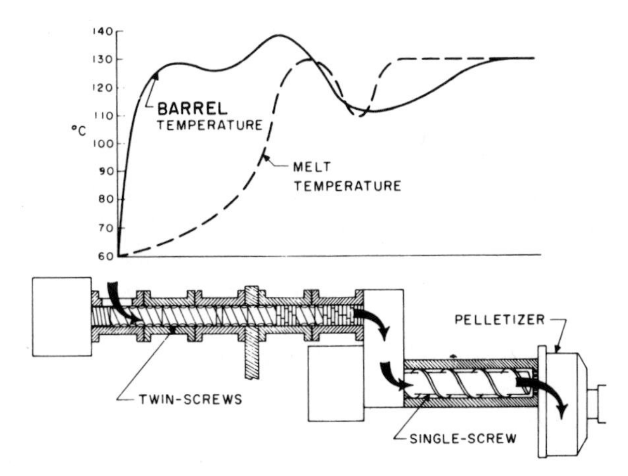

FIG. 9. Barrel and melt temperatures in a two-stage compounding system.

5.4.4 Shear Cone Compounding

Another compounding system, (Fig. 10) designed for the compounding and pelletising of flexible PVC formulations combines a twin-screw feed section (for positive feeding and therefore minimum degradation), a large rotating cone (for imparting high shear) and a single-screw extruder (for final extrusion through a pelletising die).

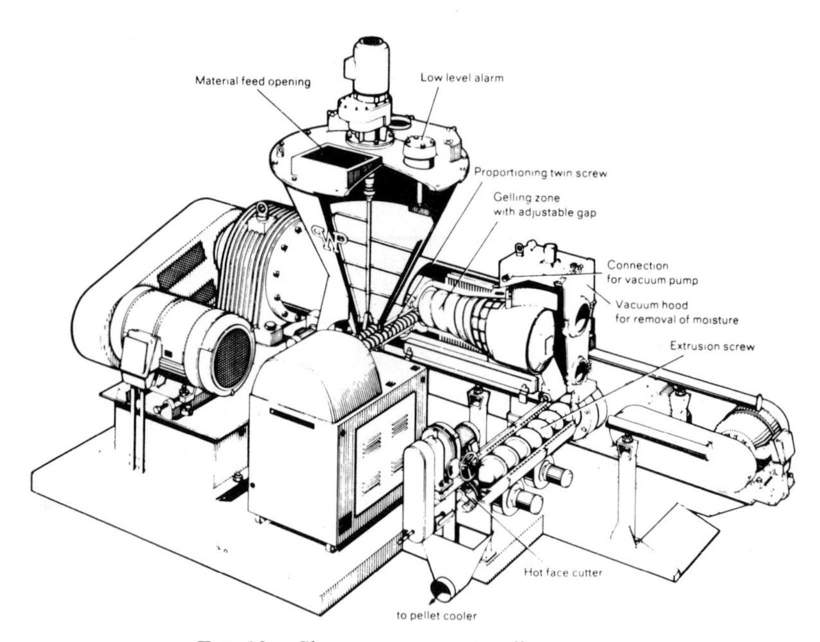

FIG. 10. Shear cone compounding system.

Material, usually in the form of pre-mix from a ribbon blender, is fed into an agitated hopper and then to a short twin-screw feed extruder for pulsation-free conveying to the shear cone. The shear cone is a ribbed conical screw that rotates in a corresponding conical housing or barrel. The diameter of the cone increases in the direction of material flow and the space between the cone and the housing can be varied from 0·3 to 6·0 mm (0·012 in to 0·24 in) during operation.

When the material is forced between the cone and housing, it is plasticised, gelled and homogenised. Because of the high shear generated by the rotating cone, no additional energy is required to soften the material.

This results in a very even temperature distribution, as the thickness of material is fairly small. Short residence times, usually less than 25 s are required in this section.

The ribs on the cone roll the material into cigar-shaped pieces which fall on to the (single) discharge screw, which then conveys them through the die and to a hot face cutter. Table 2 gives typical operating conditions of a shear cone compounding system.

TABLE 2
TYPICAL OPERATING CONDITIONS OF A SHEAR CONE COMPOUNDING SYSTEM

Main drive	*Average hourly output of plasticised PVC*	
kW	*kg*	*lb*
30	330	726
55	700	1 540

5.5 THE COMPOUNDING FACILITY

Depending on the requirements of the processor, a separate compounding and pelletising installation for PVC can take a number of different forms. Figure 11 shows the steps involved in compounding. Systems for handling and storage of resin either in bulk or in bags can be devised for economical operation, depending on the planned output rate of the compounding line and the number of different ingredients to be compounded. The same considerations also apply to handling of plasticisers, fillers, reinforcements and other additives.

Three types of storage and handling systems are generally available:

1. *Manual feed system*. This system is economical for output rates up to 1 ton/h but is limited by the fact that the resin-to-filler ratio must be on a constant bag-to-bag basis in order to avoid preweighing (Fig. 12).
2. *Semi-automatic system*. This system is economical at rates above 1 ton/h where preweighing of additives, etc. is kept to a minimum (Fig. 13).

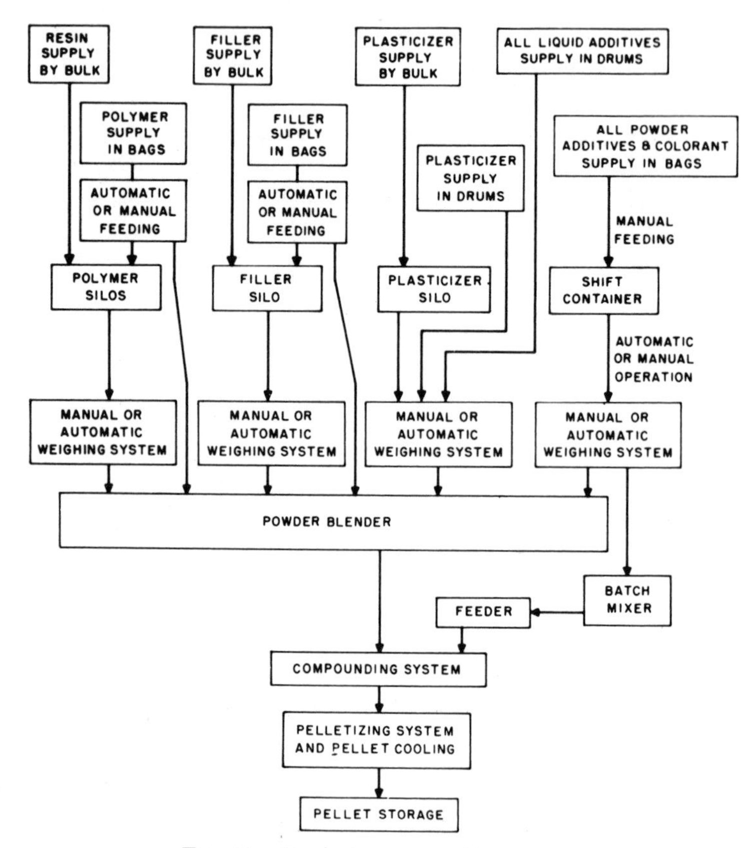

FIG. 11. Typical compounding system.

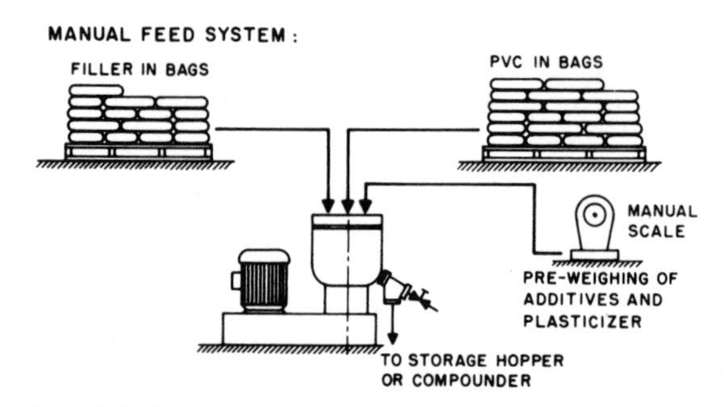

FIG. 12. Manual feed system for use with high speed mixer or ribbon blender.

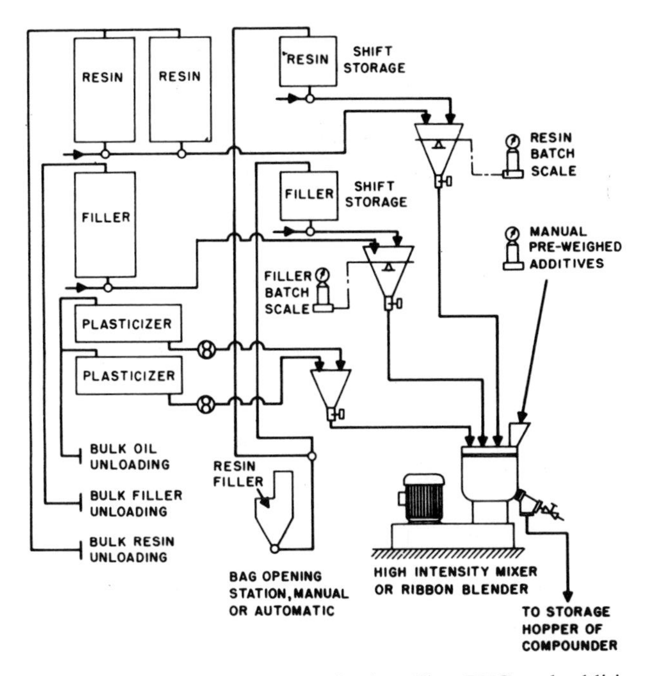

FIG. 13. Semi-automatic system for handling PVC and additives.

3. *Automatic system.* This is the most economical system for high output compounding operations. It has the advantage that stabilisers, lubricants and colourants are automatically weighed and metered into the premixer (Fig. 14). The plant can be interlocked to stop the operation in case of any malfunction or failure to dispense any preset ingredient.

For the most part, bulk storage systems are more economical for systems designed to handle more than $\frac{1}{2}$ ton/h. An automatic compounding system such as that shown in Fig. 14 can be equipped with either fully automatic feed systems or systems using bulk resin with a batch powder blender and weighing system.

With any of these systems, the critical factor in terms of economics and product quality is the type of compounding equipment selected and it is in this area that the most recent advances have been made. While single-screw extruders are still used in this intermediate compounding operation, their ability to perform the tasks of mixing, gelling, devolatilising and conveying

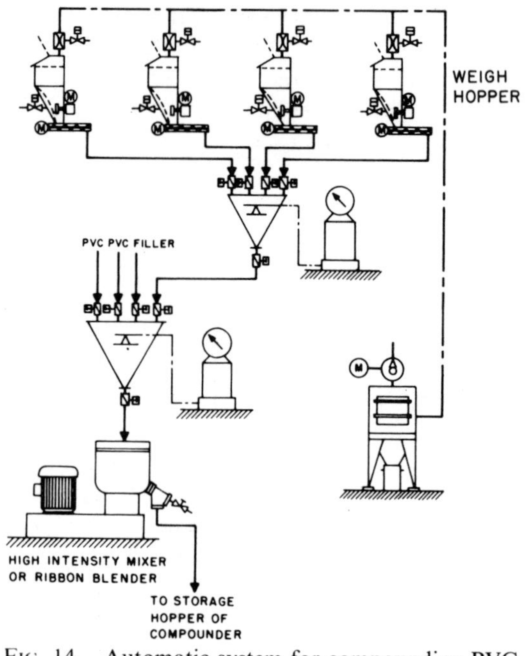

FIG. 14. Automatic system for compounding PVC.

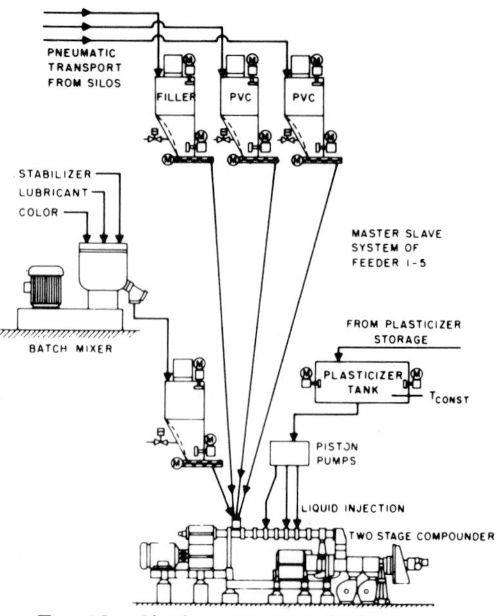

FIG. 15. Single component feed system.

in one operation is limited. Although several new single-screw configurations have been developed, they necessarily compromise some part of the operation to optimise others.

5.6 IN-PLANT COMPOUNDING

The advantages of being able to tailor PVC formulations specifically to the application and to recycle PVC scrap directly by in-plant compounding were generally thought to be beyond the capability of any but the largest processors. Today, the annual volume requirements needed to justify in-plant compounding have been considerably lowered. This is due to the changing economic situation and also to new developments in compounding equipment. Thus, processors who handle upwards of 400 tons (400 tonnes) of PVC compounds per year can often realise substantial cost advantages by establishing in-house compounding facilities.

This decision, of course, is based on the comparative investment cost, operating cost and payback. By considering the case of a 650 kg/hr shear cone compounding system for flexible PVC, the equipment requirements for in-plant compounding can be demonstrated.

The basic steps in flexible PVC compounding are preblending, compounding and pellet handling. Disregarding the feed method, a compounding system like that shown in Fig. 16 consists of a high speed mixer (although heatable ribbon mixers are also suitable), in which plasticiser is absorbed by the PVC. The resulting pre-mix is conveyed directly to the shear compounder without intermediate dry-blend cooling which reduces energy consumption by 0·04 kWh/kg and increases throughput rates. The material is gelled and pelletised in the shear cone compounder after which the pellets are cooled from about 150 °C to below 43 °C in the pellet cooler.

To avoid the necessity of continuous metering of small proportions of additives (some of which, e.g. stabiliser or pigment paste, are difficult to feed), it is customary to pre-mix the polymer powder, liquid plasticisers, fillers and all other ingredients in a batch mixer and then to feed the pre-mix either into an agitated intermediate hopper or direct into the hopper of the compounding machine (Fig. 16).

The basic plant layout can be varied to meet specific requirements for bulk storage of individual components, automatic feeding of dry-blend, automatic loading of storage bins or immediate processing after cooling.

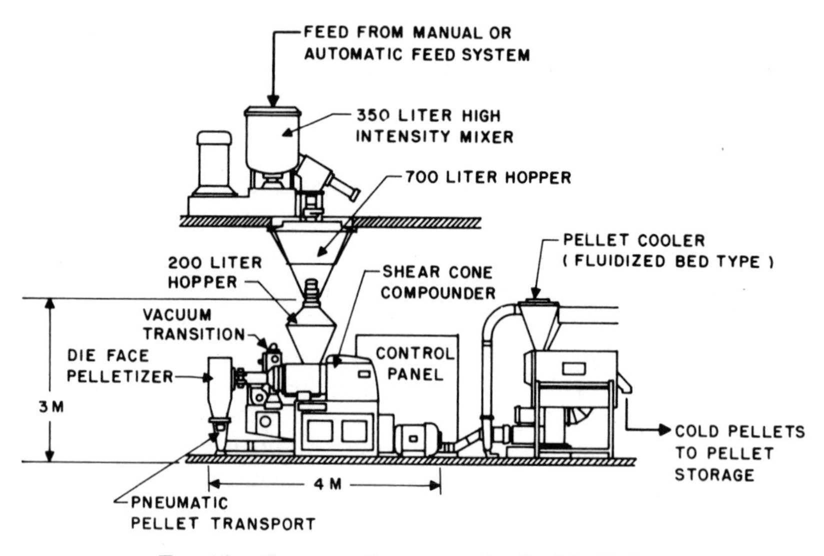

FIG. 16. Compounding system for flexible PVC.

5.7 CONCLUSION

PVC continues to be one of the world's most widely used plastics. It is highly versatile material and can be processed by virtually all the different plastics processing methods. Its product applications include every major industry, where it is used in thousands of different formulations.

The material's ability to readily accept modifications by changing both its processing characteristics and final product properties by compounding it with additives and modifiers gives it unique universal characteristics. The compounding operation, therefore, is a critical step in the base-resin to end-product development process. It directly influences both total material economics and end-product properties.

Significant advances in PVC compounding technology have occurred in recent years which have further enhanced the material's competitive position. These include the development of specialised PVC compounding systems such as the two-stage and shear cone processes which have been economically justified for relatively small as well as medium and large volume operations.

Chapter 6

CELLULAR PVC

K. T. COLLINGTON
Fisons Ltd, Cambridge, UK

SUMMARY

Polyvinylchloride is probably the polymer most widely used to produce cellular products. It has been estimated that approximately 800 000 *tonnes of PVC polymer is used for such applications throughout the world. Such cellular products are produced mainly by chemical expansion systems using blowing agents such as azodicarbonamide. Originally, cellular PVC was produced by press moulding techniques but nowadays the extrusion process and processes based on plastisols are receiving a great deal of attention. Such developments are therefore described in the following chapter.*

6.1 INTRODUCTION

Polyvinylchloride (PVC) is probably the most widely used polymer in the production of cellular materials, independent of production process or expansion system. World consumption of PVC polymer in the production of cellular materials is estimated at 6–800 000 tons per annum. The main area of application is in highly plasticised PVC, but developments in both rigid unplasticised and semi-rigid foams exist.

Production is by both physical and chemical expansion systems but the majority of foam production is based on the use of chemical expansion systems, predominantly using azodicarbonamide. Surprisingly, the inherent thermal instability of PVC, necessitating the use of metal salts as thermal stabilisers, provided the key to the use of chemical expansion systems in PVC foams. Lead, zinc and cadmium salts, commonly used as thermal stabilisers, interact with azodicarbonamide (decomposition temp. 227 °C), reducing the decomposition, or gas evolution temperature to within the conventional processing temperature range.

Cellular PVC was first produced commercially using press moulding/chemical expansion systems and used emulsion grade homopolymers, plasticised, or blended with crosslinkable monomers for the production of ultra-low density, rigid foams. Problems arising from the toxic decomposition residues of azodiisobutyronitrile, used as a source of both gas and free radicals in the case of rigid foams, has severely restricted development in this area of foam production.

Concurrent with the development of press moulded foams, development of cellular-coated fabrics and unsupported cellular sheet began, resulting in the formation of the main area of application for cellular PVC existing today. Calendering techniques are used for the production of cellular coated fabrics, but most production is based on plastisol coating systems. This production technique, based on emulsion grade homopolymers, is a logical extension of the existing fabric coating industry and offers advantages in lower production costs and flexible production schedules.

Formulation development in this area has resulted in the production of supported and unsupported sheet with fine, uniform cell structures of low density. Improvements in the abrasion resistance and handle of coated fabrics are a result of developments in skin formulations based on polyurethane and polyimide surface coatings. Development in the production of 'breathable' cellular coated fabrics, has not been entirely successful and end-applications are mainly restricted to shoe construction materials. Processes are based on careful control of melt rheology and gas evolution rate to give open cells in chemical expansion systems: leaching techniques and/or multi-lamination systems can be used to achieve the necessary porosity.

The most recent developments in the coating area are the introduction of novel, patented, chemical embossing techniques,[1,2] to achieve sculptured surface textures on coated substrates for use in wall and floor coverings.

Development in the production of thick section, plasticised PVC foams, produced by gas solubility techniques,[3] suitable for use in automotive and upholstery applications, is restricted to process and end use application. In some applications these materials are being displaced by crosslinked polyolefin foams of lower density and improved physical properties.

The injection moulding and extrusion of cellular PVC are currently the subjects of considerable development (particularly the extrusion of cellular unplasticised pipe and profile). Development of extruded cellular unplasticised profiles (as wood replacements in the construction industry) has been rapid, particularly in North America. Developments in Western Europe are apparent in both cellular pipe and in thick, complex section profiles suitable for use in the building industry.[13]

The injection moulding of cellular PVC is restricted mainly to blends, based on plasticised homopolymers and nitrile rubber or urethane/PVC blends, which find wide application in the shoe industry. Development in unplasticised homopolymers and ABS/PVC blends will remain embryonic until processing and thermal stability problems are resolved.

Other miscellaneous development areas exist, but the information available is restricted and refers mainly to process development, e.g. nitrile rubber/PVC blends, for the production of pipe insulation using pre-forming, free expansion techniques, cellular gaskets produced by screen printing techniques, bottle seals and other packaging applications.

6.2 EXPANSION SYSTEMS

Both physical and chemical expansion are used in the production of cellular materials, but the majority of production processes are based on chemical expansion systems.

6.2.1 Chemical Expansion Systems

Chemical blowing agents evaluated for the production of PVC foams include benzene sulphonhydrazide, p′p″ oxybis benzene sulphonhydrazide (OBSH), dinitroso compounds and azodiisobutyronitrile (AZDN). However, the most widely used chemical blowing agent is azodicarbonamide. The reasons for using azodicarbonamide are that:

1. The original material, the gaseous decomposition products and the residual solid decomposition products are non-toxic, thus satisfying Food and Drug Administration (FDA) regulations;
2. The gas yield is high—220 ml/g at 220 °C;
3. Both the rate and temperature of decomposition are controllable by the selection of activator/stabiliser system and the particle size of the azodicarbonamide.

Numerous technical papers are available outlining the performance requirements of a chemical blowing agent with particular reference to azodicarbonamide.[4,5]

Developments in chemical expansion systems are mainly aimed at improving the performance of azodicarbonamide and are based on two inter-related areas.

1. *Control of the Particle Size of Azodicarbonamide*

Numerous grades of azodicarbonamide are available with a wide range of particle sizes (average particle diameter 2·0 μm–20·0 μm) and of differing particle size distribution. Such factors are dependent on the method of production. Production of this range of powders is normally by simultaneous grinding and classification in fluid energy mills, wherein the air/azodicarbonamide ratio is high thus eliminating any problem of pre-decomposition due to localised heat build-up.

The effects of differing particle size distributions on the decomposition rate are illustrated in Fig. 1.

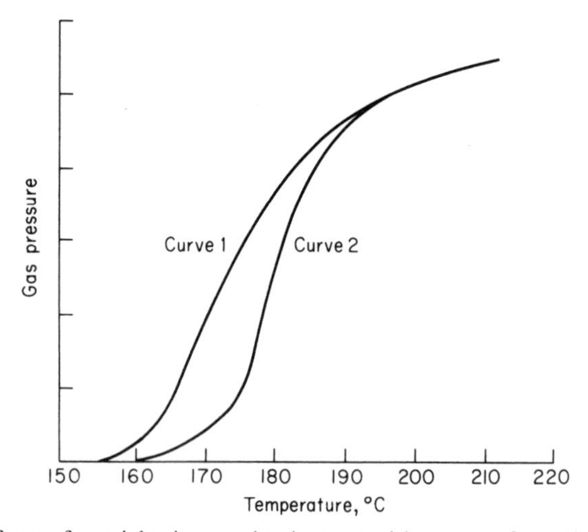

FIG. 1. Effects of particle size on the decomposition rate of azodicarbonamide. *Curve* 1: Average particle diameter = 4 μm; *Curve* 2: Average particle diameter = 10 μm.

Since the decomposition rate of azodicarbonamide alone is autocatalytic (due to the liberation of urea, a known activator for azodicarbonamide), the effects of changes in particle size are significant. A predominance of fine particles with high surface area/mass ratios decompose rapidly liberating urea, speeding up the overall rate of decomposition of the powder. These subsequent variations in the rate of decomposition of the azodicarbonamide, together with the selection of the activator/stabiliser system, necessitate modifications to formulations to achieve the hot melt strength necessary to retain the evolved gas as finite cells within the polymer melt.

2. *Activation of Azodicarbonamide by Thermal Stabilisers*

A number of commonly used thermal stabilisers reduce the decomposition temperature of azodicarbonamide. They include the salts, organic complexes, or co-precipitated soaps of lead, zinc, and cadmium. The degree of efficiency of activation of these metallic complexes is illustrated in Fig. 2 using the same particle size azodicarbonamide.

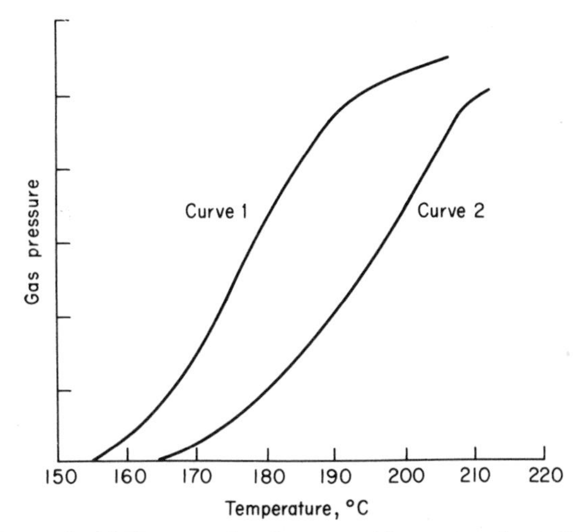

FIG. 2. Effects of stabiliser on the decomposition rate of azodicarbonamide, average particle diameter 10 μm. *Curve* 1: Zn/Cd stabiliser; *Curve* 2: Ba/Cd stabiliser.

Comparison of Figs. 1 and 2 indicates the degree of control attainable over both the temperature of decomposition (of the azodicarbonamide) and the rate of gas evolution, thus permitting freedom in formulation which is necessary to obtain other physical properties in the cellular material.

Current developments aimed at modifying the performance of azodicarbonamide include the production of pastes containing activators, the deposition on the surface of the azodicarbonamide crystal of salts of known activators,[18] the addition of small quantities of colloidal silica, and the addition of small quantities of other blowing agents.[19]

6.2.2 Physical Expansion Systems

Development in this area is at a low level and is based entirely on plastisols using either siloxanes or incompatible soaps, in air entrainment techniques;[9] or increasing the solubility of gases in a plastisol by adjustments of temperature and pressure.[3]

The use of cell stabilisers in conjunction with chemical expansion systems has been evaluated as a means of preventing cell collapse in plastisols during fusion, but results appear inconclusive.[7] Developments in unplasticised extrusion do however indicate that cell stabilisers used in conjunction with other melt modifiers (acrylics) can improve cell quality and surface appearance.[8]

However, in conclusion most of the development work currently being undertaken is based on chemical expansion systems involving the use of azodicarbonamide in one of the forms mentioned above.

6.3 PLASTISOL DEVELOPMENTS

Cellular coated fabrics, wall coverings and floor coverings, based on asbestos paper and other substrates, constitute the main areas of application for expandable plastisols.

6.3.1 Production Techniques

The production techniques used vary according to the type of product. Transfer coating techniques are used in the production of cellular coated fabrics based on knitted fabric backings. Lamination of the knitted fabric backing is on to the gelled or liquid plastisol formulated to minimise 'strike through' or 'wicking' into the fabric during the expansion stage. The use of pre-embossed release papers (as a replacement for smooth release papers or stainless steel belts) is increasingly replacing subsequent mechanical embossing systems. The use of skin casting techniques, utilising pre-embossed release paper, increases the emboss retention temperature thus satisfying certain automobile service specifications. High gloss, fine grain finishes suitable for use in fashion goods are also produced by this casting technique.

Coated fabrics based on woven and non-woven fabrics are produced by direct coating of the expansion layer and the top coat, or wear layer, on to the filled calendered fabric. Similar techniques of direct coating are utilised in the production of wall and floor coverings. Both reverse roll coating systems and conventional knife over steel roll coating techniques are used in the production of wall coverings and floor coverings.

Melt coating techniques (the subject of considerable development work) are not used extensively in commercial production although an extrusion coating/lamination technique based on suspension polymers has been used in small scale production. Obvious advantages of melt coating techniques are savings in raw materials costs, i.e. the use of suspension grade polymers, and reductions in plasticiser levels—of particular interest in automotive 'anti-fogging' interior trim applications.

6.3.2 Formulation Development

Little development is apparent in this area, with formulations still based almost entirely on phthalate type plasticiser systems. The use of polymeric or other speciality plasticisers is restricted to the production of speciality foams.

Developments are aimed at increasing output rates and involve the use of stabiliser/activator systems and fine grades of azodicarbonamide as described previously. Considerable work has been published on the subject of activation of azodicarbonamide together with the effects of formulation variables.[9] Rheological studies[10,12] clearly indicate the critical relationship between polymer fusion temperature and gas evolution temperature, relative to the quality of cell structure.

6.3.3 Process Development

Process development, particularly in the production of cellular wall coverings and floor coverings, has been most significant. Two patented processes[1,2] are currently being used commercially in the production of chemically embossed, supported and unsupported sheet. Both processes are based on the same concept, i.e. the rates of expansion in selected areas of an expandable sheet are capable of being controlled to enhance multi-coloured printed surfaces and thus obtain sculptured, three dimensional effects.

The major difference between the two processes is the method used to control the expansion in selected areas. In one process[2] the rate of expansion in a gelled sheet (containing azodicarbonamide) is enhanced in selected areas by the printing of formulated inks containing activators for the azodicarbonamide on to the surface prior to the application of the transparent wear, or print protection layer and subsequent expansion.

In the second process,[1] the procedure is similar but the printing ink contains additives which inhibit the decomposition of the azodicarbonamide, i.e. the decomposition temperature of the azodicarbonamide in the printed areas is increased relative to that of the azodicarbonamide in the adjacent unprinted areas, resulting in differential expansion in the desired pattern.

Figures 3 and 4 show the activation and inhibition processes, respectively. These figures illustrate the differences obtained in the two processes after expansion in a circulating hot air oven at an air temperature of 210 °C.

Advantages of these processes include:

1. the elimination of problems of register (between multi-colour printing units and mechanical embossing units) and
2. a saving in machinery costs (due to the elimination of an embossing unit and the range of shell patterns necessary).

Either process may be used for the production of cellular floor or wall covering, but the 'inhibition process' is most widely used in the production

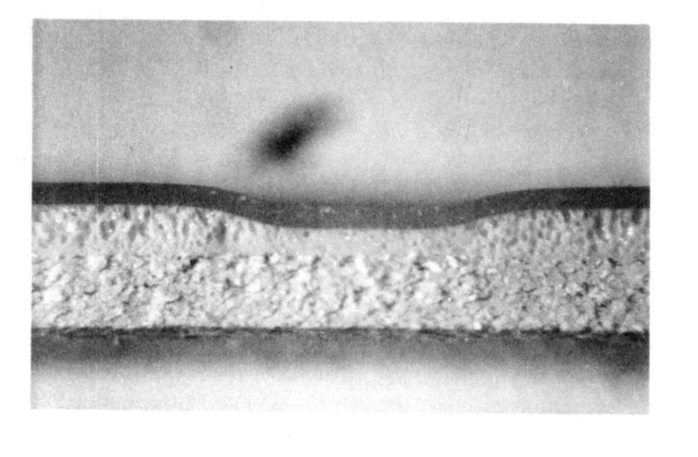

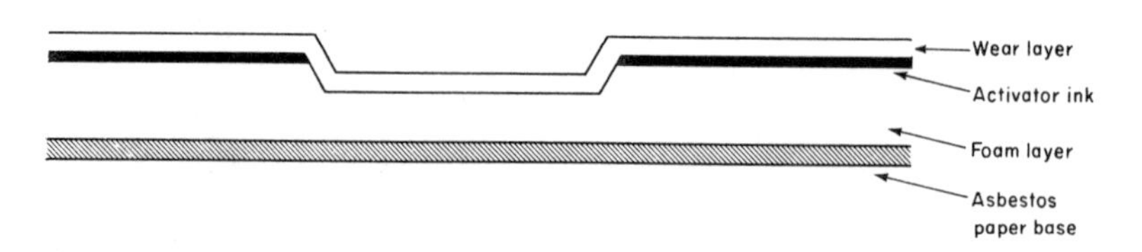

Fig. 3. Activation process.

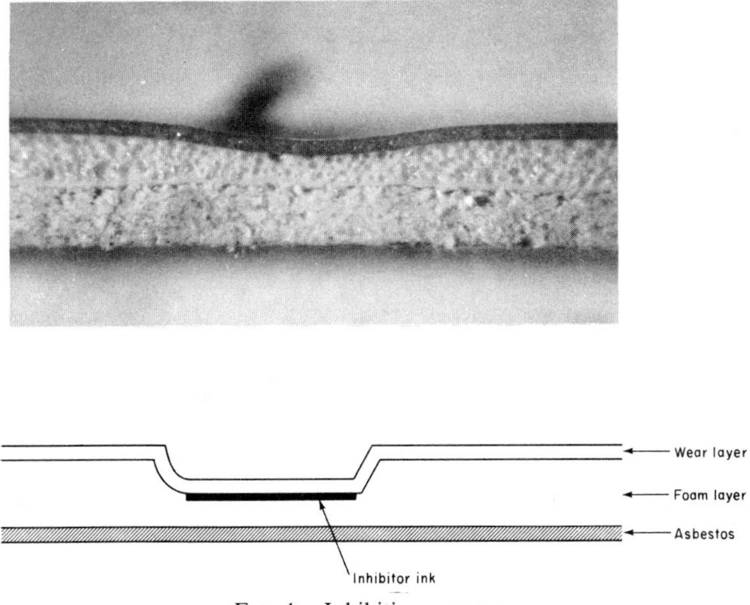

FIG. 4. Inhibition process.

of floor coverings and the 'activation process' is used in cellular wall covering applications.

A further process, based on the printing of plastisols containing blowing agents[12] so as to obtain sculptured effects, is also being operated. A mechanical embossing process utilising extremely deep embossing rolls which completely crush the foam in selected areas so as to obtain excellent definition of a pattern, is also currently being operated in Japan.

6.4 PRESSURE BLOWN FOAM

Developments in the production of flexible and rigid closed cell foams, where both fusion and decomposition of the blowing agents occur within the mould cavity, are limited.

Formulation developments are restricted to the evaluation of chemical expansion systems designed to replace the AZDN/isocyanate systems used in rigid foams and the replacement of OBSH used in flexible foams. Attempts to replace both the above expansion systems with azodicarbonamide are, as yet, not completely successful. Problems with polymer degradations, as a result of the exothermic decomposition of azodicarbonamide at levels of addition of 10–14 phr, have not yet been overcome. Activated grades of azodicarbonamide of selected particle size are, however, used successfully in the production of flexible foams.[17]

Process development in this general area is not advanced since a two-stage process is involved. A high clamping force press is necessary to ensure the solubility of the evolved gas in the melt (during both fusion and cooling cycles). Expansion in a steam chest or a silicone bath, in the case of rigid foams, then occurs. Development in this area, aimed at reducing the current toxicity problems arising with the AZDN/isocyanate expansion systems, can be anticipated.

This method of production is used for fishing floats, suitable for use in deep sea trawling, and is currently the only method available. The possibility of using modified injection moulding techniques is being investigated but poses major problems in mould design as a volume expansion factor of $\times 10$ is necessary to achieve the density reductions required.

In conclusion, development work on pressure blown foams will be concentrated on rigid foams only, since it is doubtful whether flexible foams produced by this process can compete economically with either urethanes or crosslinked polyolefins. Potential areas of application for these materials include flotation jackets and other buoyancy applications, vehicle insulation, sports goods and building applications.

6.5 THE EXTRUSION OF CELLULAR PVC

The extrusion of cellular, unplasticised PVC pipe and profiles has been the subject of extensive development since 1970. Development is concentrated on two types of extrusion process:

1. Free expansion systems, where expansion of the extrudate occurs beyond the main die, prior to entering the vacuum shaping and sizing units; and
2. Controlled expansion systems,[13] where expansion of the extrudate is controlled by extensions of the main die. Additional to the die/shaper extensions, a mandrel of decreasing diameter projects through the main die into the shaping area thus controlling the rate of foaming by controlling melt pressure decay within the water cooled shaping dies.

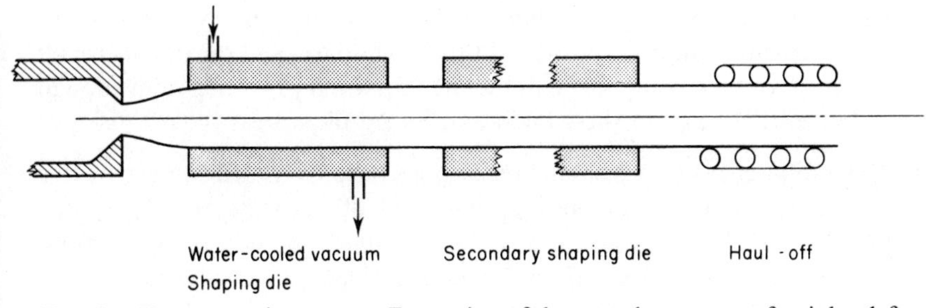

FIG. 5. Free expansion system. Expansion of the extrudate occurs after it has left the main die.

The essential differences between the systems illustrated in Figs. 5 and 6 are:

1. In free expansion systems the extrudate expands to fill the shaping die and as a result it is difficult to obtain extrudates of low density with smooth homogenous outer skins;
2. In controlled expansion systems, the rate of decay, and hence foaming of the melt, is controlled in proportion to the rate of increase of cross-sectional area within the shaping unit. The effect of controlling the rate of pressure decay, within the melt, is to collapse the foam at the interface of the shaping die extension so as to create thick, smooth, homogenous outer skins. Low density cores are obtained since foaming occurs inwards as the diameter of the mandrel decreases; and

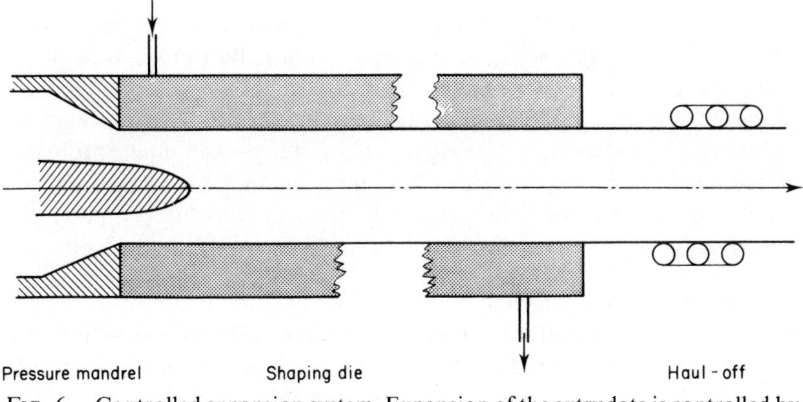

FIG. 6. Controlled expansion system. Expansion of the extrudate is controlled by main die extensions and projecting mandrel.

3. In free expansion systems, the high melt pressure required, to ensure complete solubility of the evolved gases in the polymer melt prior to passing through the die, has given problems in die design particularly on sheet and complex profiles.

The differences in processing techniques have restricted the use of free expansion systems to the production of extrudates of 1–3 in^2 (645–1935 mm^2) cross section, whilst complex extrudates with cross-sectional areas of 5–20 in^2 (3226–12 903 mm^2) have been produced using the controlled expansion techniques.

6.5.1 Equipment Considerations

Production of cellular extrudates is done on both vertical and conventional horizontal single-screw extruders, with twin-screw extruders used mainly for cellular pipe production. No changes in screw geometry other than for ensuring uniformity of heat transfer have been made, LD ratios are 20:1 plus, with compression ratios of 2:1 and 2·5:1 being commonly used.

The main areas of development are in the design of dies and calibration dies in the case of free expansion systems. The major problem in die design is one of ensuring uniformity of melt pressure across complex cross sections so as to prevent pre-foaming and loss of surface smoothness. The problem of controlling melt pressure is most pronounced in the production of sheet using conventional, centre-fed dies, where wide differences in melt pressure between the centre and the extremities of the flow paths can occur. Small scale production of plasticised cellular sheet for use in coated fabric production is known, but no unplasticised sheet production is apparent at present for the reasons outlined above.

Both dry vacuum calibration and water lubricated calibration systems are used. The length of the sizing unit is normally double that of an homogenous line. The use of two consecutive calibration dies is necessary to increase heat extraction and thus set the surface. Setting must be sufficient to overcome the internal gas pressure and thus eliminate problems of post-expansion. A typical downstream system for cellular rigid pipe or profile would consist of a vacuum sizing unit, a cooling tank containing a submerged sizing die, and finally a conventional spray cooling unit.

6.5.2 Formulation Variables

The majority of unplasticised 'free expansion' cellular extrudates are produced from dry powder blends although granular compounds are available. No modifications to blending techniques are necessary in the production of such dry blends, other than a careful control of temperature during and after the addition of the chemical blowing agent. Normally, the drop temperature of blends is restricted to 100–120 °C to eliminate possible pre-decomposition of the chemical blowing agent.

Basic requirements of a formulation suitable for use in the production of unplasticised cellular extrudates are:

1. A high level of hot melt strength, but a relatively low melt viscosity to create the rheological conditions necessary for cell formation; and
2. Rapid set up characteristics of the melt so as to permit the formation of a smooth, high density outer skin.

The above factors are pertinent to both expansion systems, but are more important in 'free expansion' systems where the control of skin formation is more difficult.

Formulation development aimed at achieving the above rheological requirements has involved the evaluation of the compounding ingredients.

The resins commonly used are K value (55–60) homopolymers as these give good melt strength. Developments in North America involve the use of olefin copolymers as both the sole polymer, or as a partial replacement for the homopolymer; the material then acts as a melt modifier.[20]

Melt modifiers are essential to attain the necessary hot melt strength. The majority of free expansion formulations contain high levels of acrylic processing aids, in preference to ABS, SAN and olefinic melt modifiers which have lower efficiency, as acrylics give better and more uniform cell structures.

Lubricants are essentially selected relevant to the extruder type, i.e. single- or twin-screw. They are also of value, however, in increasing gelation rates thus ensuring complete fusion and melt homogeneity prior to decomposition of the blowing agent. In free expansion systems the levels of internal lubricants are high. Levels of external lubricants, kept to a minimum in free expansion systems, are increased in controlled expansion systems to enhance skin formation.

Thermal stabilisers have a dual function, acting as activators for the blowing agent (mainly azodicarbonamide) and as heat stabilisers. Their selection will be related to other formulation variables, ensuring that decomposition of the blowing agent occurs after fusion so as to permit solution of the evolved gas in the melt. Both lead and barium/cadmium type stabilisers are used, but if optimum thermal stability, without toxicity problems is required, modified tin stabilisers which act as activators for azodicarbonamide are used.[15]

The chemical blowing agents mainly used are sodium bicarbonate in the controlled expansion system and azodicarbonamide, either alone, or blended with nucleator systems in free expansion systems. Development work in the controlled expansion system is aimed at replacing the sodium bicarbonate with an organic system so as to eliminate staining problems and the reported degradation of physical properties on ageing. In free expansion systems, organic expansion systems decomposing at 145–160 °C, with improved cell forming characteristics, are being developed based mainly on modified azodicarbonamide systems.[18]

6.5.3 Process Development

The process development in 'free expansion' systems is mainly aimed at improving surface hardness of the extrudate so as to give physical properties comparable with those obtained using controlled expansion techniques. Annealing of the surface is a system currently being evaluated although the use of co-extrusion systems recently developed for ABS pipe[16] could be applicable to unplasticised PVC.

In conclusion, further developments in the extrusion of cellular unplasticised profiles can be anticipated using the controlled expansion systems, where the production of complex profiles is more viable than in wood or aluminium. Free expansion systems will continue to be used mainly for the production of small cross-section extrudates only, e.g. skirting boards, architraves, and sidings. A possible exception is in the production of cellular non-pressure pipe and conduit, currently being produced commercially in France using conventional twin-screw equipment.[22]

6.6 INJECTION MOULDED CELLULAR PRODUCTS

Production in this area is mainly in plasticised homopolymers, nitrile or urethane/PVC blends. The end applications are predominantly in the light shoe industry.

Developments in this area are static with competition from thermoplastic rubbers and EVA compounds increasing, particularly in the fashion shoe markets.

The moulding of unplasticised PVC foams is embryonic and development is currently at the laboratory experimental moulding stage. Problems with thermal stability, common to the moulding of homogeneous unplasticised PVC, are magnified in cellular moulding operations. This is due to long residence times in the accumulator and shear heating caused by the high injection speeds used on foam moulding equipment. Trials carried out by polymer producers in W. Germany are reported to be successful, but no commercial production is yet apparent.

6.7 MISCELLANEOUS DEVELOPMENTS

Rotational casting of toys and automotive interior trim has been investigated, but the long cycle times necessary for the production of integral skin foam mouldings, together with problems of thermal stability, have restricted development. Other plastisol developments include 'crown cork' materials, widely used in soft drink cap liners, and 'cook-in-the-pack foods'. Formulations are based on either activated azodicarbonamide, or OBSH, and certain micro-crystalline waxes (to prevent adhesion between the cellular cap liner and the glass). Packaging systems involving both expandable, plasticised powder blends and plastisols have been developed in the USA but none are of commercial significance.

REFERENCES

1. British Patent 106998 to Congoleum Ind., USA.
2. British Patent 1147983 to Fisons Ltd, London.
3. US Patent 2666036 to Elastomer Chemical Corp., Newark, USA.
4. Reed, R. A. Chemistry of modern blowing agents, *Trans. British Plastics Conv.*, 1955.
5. Lasman, R. L. *Encyclopedia of Polymer Science & Technology*, Vol. 2, 1965, pp. 532–65.
6. Patents to Vanderbilt Corp., USA.
7. Air Products & Chemicals Inc., Philadelphia, USA, Technical Bulletin VS 103.
8. Rohm & Haas Inc., Philadelphia, USA; *Resin Review XXI*—Acryloid K1479120N.
9. Lally, R. E. and Alter, L. M. Activator performance in vinyls, *J. Soc. Plast. Eng*, Nov. 1967.
10. Nass, L. I. Compounding of low density vinyl foams, *Modern Plastics*, March/April 1963 (2 parts).
11. Hodgson, T. C. *et al.* An instrument for the analytical control of the chemical blowing of polymers, *J. of Cellular Plastics*, Nov./Dec. 1973, pp. 274–8.
12. Patented Process by Stork Brabant BV, Holland.
13. British Patent 1184688 to Produits Chimiques, Ugine Kuhlmann, Paris.
14. *Plastics Technology*, Feb. 1973, pp. 42–4.

15. DWORKIN, R. D. Decomposition of azodicarbonamide in the presence of organotin stabilisers, *Trans. 35th SPE ANTEC*, Montreal, April 1977.
16. Non co-extrude foam core ABS pipe, *Plastics Technology*, Dec. 1976, pp. 50–2.
17. German Patent 2457977, Lonza, Basle, Switzerland.
18. Japanese Patent 0087455, Eiwa Chemical Co., Kyoto, Japan.
19. US Patent 3743605, Uniroyal Inc., USA.
20. German Patent 2000039, BASF Ludwigshafen, W. Germany.
21. No pipe-dream, *Chemical Week*, 23 Feb., 1977, pp. 32–3.
22. Tech. Lit. Ref. N172E, Armosig; Celle, St. Coude, France.

Chapter 7

BLOW MOULDING

J. PICKERING

Fibrenyle Ltd, Beccles, Suffolk, UK

SUMMARY

The technique of extrusion blow moulding of PVC took many years to develop, but expanded rapidly in the late 1960*s and early* 1970*s, following the development of materials and blow moulding machines.*

A large part of the total blow moulded market for general packaging is now in PVC and PVC competes well with other thermoplastics in a wide range of markets.

This chapter attempts to set out some of the critical processing characteristics for PVC and the developments which have led to successful blow moulding of PVC bottles. A brief look is also taken at future trends and recent developments in this area.

7.1 INTRODUCTION

Since the earliest days of blow moulding PVC has been an attractive material for bottles and containers, but it suffered from inherent heat stability problems which initially retarded its growth in this market. As a replacement for glass, it offered high clarity and gloss without the heavy weight and breakage problems associated with glass.

PVC packaging of toiletry products used in bathrooms and kitchens, where broken glass is hazardous, has special attractions and now the majority of shampoos, bath oil and hair aids are packed in PVC bottles and containers.

Apart from its clarity and gloss, PVC offers superior resistance to gas permeability relative to polyethylene. Polyethylene was of course the most common plastics material initially in use for bottles and containers. The low gas permeability of PVC opened up new markets for this material, markets such as packaging for cooking oils, fruit squashes and concentrates, cosmetics, disinfectants, in fact wherever oxygen (from the air) had a deleterious effect on the contents. Although still not as good as glass in this respect, the shelf life of these products when packaged in PVC, was extended to an acceptable level.

To the blow moulder, PVC offered another interesting and attractive advantage. Because of the relatively high rigidity of PVC, bottles with thinner walls were possible. These bottles with thinner walls could be cooled more quickly than the polyethylene bottles, and hence machine output rates could be increased. PVC bottles are now therefore used to package food products, pharmaceuticals, household chemicals, cosmetics, toiletries and many industrial products.

Against all this, however, there is the serious defect of poor heat stability. Even today this continues to be an obstacle to the usage of PVC as a blow moulding material.

In the early days of blow moulding, machines were built to handle polyethylene, which is, from the point of view of the processor, one of the most heat-stable plastics materials. It should be no surprise, then, to learn that it has taken many years of development by machine manufacturers, material suppliers and processors to give acceptable run lengths when using PVC.

The first attempts to overcome this problem were made by separating the extrusions and blow moulding stages as, for example, in the Marrick System.

7.2 THE MARRICK SYSTEM

In this process lengths of PVC tubing were extruded, cooled and cut into short pieces from which bottles were later blown in a blow moulding unit. The tubes had to be re-heated to the processing temperature and this was done in ovens. Large ovens were therefore needed alongside the blow moulding units and accurate allowances were made for heat shrinkage, swell, etc. Critical temperature control of the ovens was also required.

Economically, this system was doomed to failure once the problems of handling PVC in a single step process were overcome, and this is exactly what happened. The process did, however, temporarily fill a gap in the supply of PVC bottles, and had the advantage that multi-cavity blow moulding could be used to give relatively high output rates for that time.

Now that the advantages of PVC have been recognised, machine manufacturers, PVC suppliers, and processors have all put considerable effort into developing the blow moulding of PVC and this effort has resulted in significant improvements in extrusion blow moulding. This process can be split into two steps or stages; these are:

1. extrusion of the parison(s); and
2. blow moulding.

7.3 PARISON PRODUCTION BY EXTRUSION

7.3.1 Single-screw Extruders

Extruders have mostly evolved from the early single-screw extruder used for polythene, and they have a length/diameter (L/D) ratio of about 20:1. Constant pitch, decreasing depth flights are generally used for PVC, and the compression ratios used are lower than those used for polyethylene (generally in the range 1·8:1 to 2·3:1). Because of the acidic nature of PVC degradation products, corrosion-resistant steels are generally used for screw construction. The screws may be nitrided, and are often hard chromium plated.

Provision is usually made for cooling both the barrel and the screw, as the frictional heat build-up is much higher with PVC than with polyethylene. In the case of smaller size extruders, this is generally achieved by the use of cooling air; for larger machines (above 60 mm screw diameter) oil cooling is more effective and gives better control. Obviously, temperature control of the melt is most critical with PVC, and since much of the heat generated is by friction, control of the cooling medium is as critical as the external barrel heating. This is often overlooked, with disastrous consequences.

Due to the corrosive nature of PVC degradation products and the high friction generated in the screw and barrel, wear of these parts is much more rapid than when polyethylene is being processed. The result of this is not merely a reduction in output, which can often be accepted with polyethylene, but a tendency to overheat and burn the PVC during extrusion. Fit of the barrel and screw is therefore much more critical than with polyethylene. A practical limit, to the amount of wear that can be tolerated, is reached when the clearance exceeds 0·20 to 0·25 mm (0·009 in to 0·010 in) over and above the normal tolerances.

7.3.2 Twin-screw Extruders

Twin-screw extruders are claimed to provide better mixing of PVC with more positive forward flow at lower temperatures than can be achieved with single-screw machines. However, early machines failed due to inadequate thrust bearings (because of the short centres between screws) and low powered drives. These problems have been overcome by using tapered screw designs (which allow greater spacing at the rear for larger thrust bearings) and more powerful drive units. Furthermore, venting of the barrel is more easily achieved and twin-screw machines are capable of giving a good homogeneous PVC melt at consistent and low temperatures.

These machines have been in common use for PVC pipe extrusion for many years, but their high output and cost, relative to single-screw machines, has delayed their use for blow moulding machines.

7.3.3 Cross-heads and Torpedoes

Once the PVC has left the end of the screw (at a temperature of around 180–200 °C) it has to be formed into a tubular parison. This is usually extruded vertically downwards.

Most designs of PVC head are now of the axial flow cross-head type with torpedo and spider support. A typical design is shown in Fig. 1.

A swan neck design of cross-head diverts the material through 90° at constant flow rate, then a streamlined core, or torpedo, with spider support divides the material to form a tubular flow into the die. The spider supports must be strong enough to prevent deflection of the core during extrusion, and at least one such support must be cored out to provide support air into the parison during extrusion.

The streamlining of these spiders is most important and the surface finish of all internal parts of the head must be highly polished. Non-corrosive steels should be used throughout to prevent problems, and hardening is advisable to prevent damage due to constant handling and cleaning of the parts. The die-head should be constructed from the minimum number of parts as this reduces the number of joints. A minimum number of joints between parts is also advisable, as each edge will eventually become slightly rounded, providing a potential hold-up point which starts decomposition of the PVC. If the edges of the spider become damaged, flow around the spiders is impaired, which in practice is the biggest obstacle to forming a smooth parison.

Cross-head and head are generally held at lower temperatures than the melt, but too great a difference will create uneven flow through the spider.

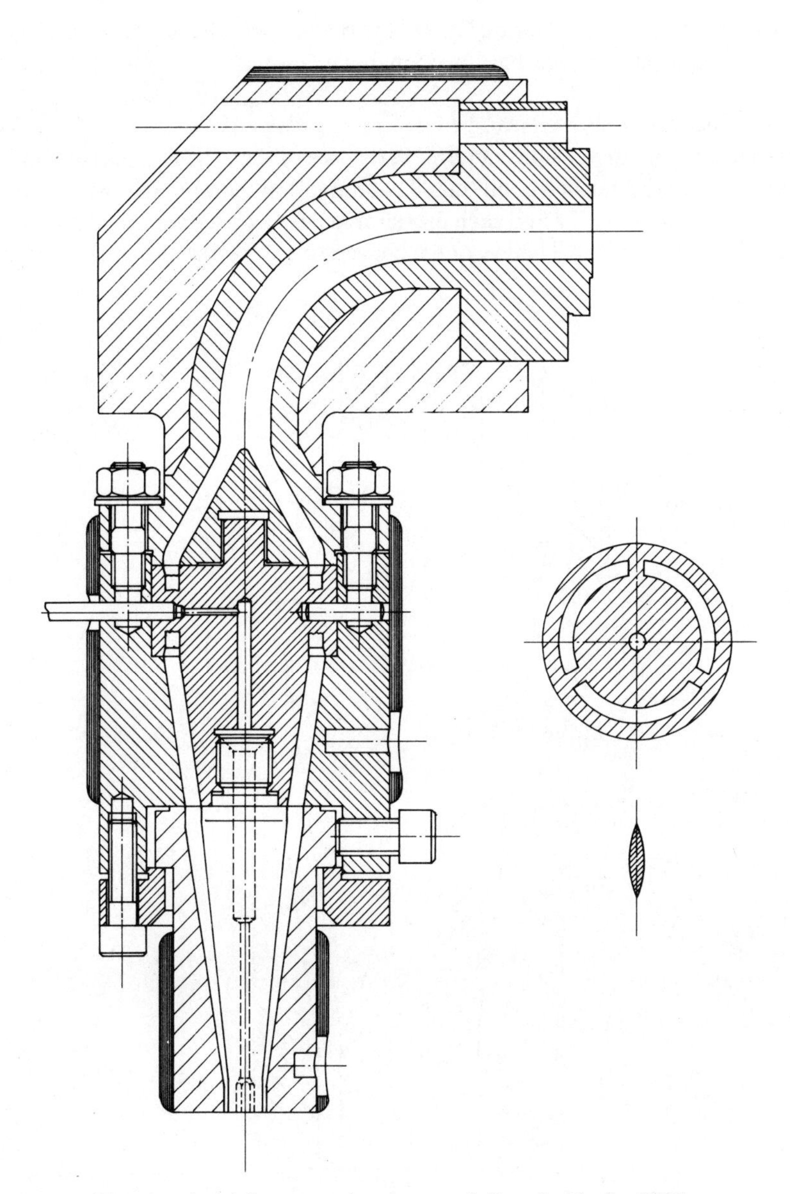

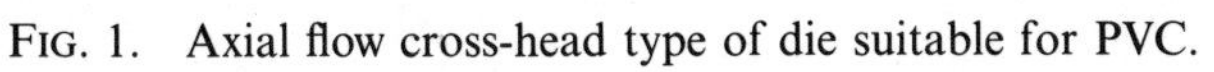

FIG. 1. Axial flow cross-head type of die suitable for PVC.

Flow lines will appear (drape effect) in the parison and the bottles produced from such a parison will have a poor finish.

7.3.4 Dies and Pins

The die and pin dimensions determine the size of the extruded parison and the weight of the finished container. The simplest of these are parallel dies of the type shown in Fig. 2 and such dies should have a lead-in angle no greater than 20 °, and a parallel length of die between 5 to 10 times the die gap. The

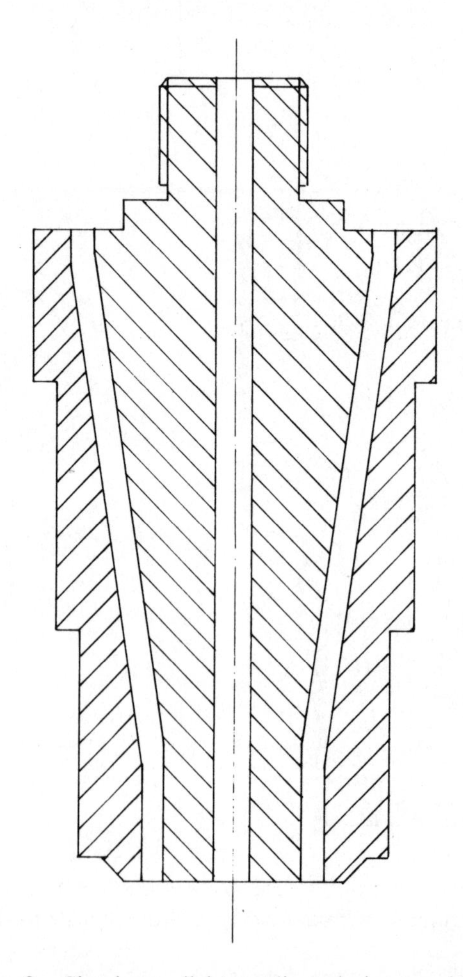

FIG. 2. Simple parallel type die and pin assembly.

die and pin should finish flush with one another so as to prevent the parison from curling and sticking to the die.

Incorrect die and pin tip temperatures can also cause curling. The temperature of the die should not be too low, relative to the melt, or curling outwards will result due to uneven flow; inward curling can occur if the tip temperature is too high.

7.3.5 Variable Aperture Dies and Parison Control

The use of dies and pins of conical shape permits variations in the die gap to be achieved by moving one relative to the other. This gives continuous control over the thickness of the extruded parison so that by appropriate programming thick and thin sections can be produced in it. Better wall thickness distribution can therefore be achieved in the bottle walls. Only slight vertical movements of the pin relative to the die are needed to give quite wide thickness variations in the parison, so accurate control of the moving part is essential.

In the case of PVC it is usual to move the die relative to the pin as this enables the early design of torpedo type head to be utilised (Fig. 3). A good bearing surface with minimum clearance is needed. With such pin and die assemblies this prevents distortions in flow and avoids decomposition of the PVC by reducing possible pick-up points.

Various mechanical and hydraulic systems have been devised to control the vertical movement of the pin, and the latest systems use very sophisticated solid state, multipoint electronic controls which can be programmed to suit the variations in parison thickness needed to give the even wall thickness in the final bottle. The firms of Hunkar, Bekum and Moog, make parison programming systems.

An advantage of these systems is that easy control of bottle weight is possible by adjusting the die gap during extrusion. It should be noted, however, that with good bottle design it is not necessary to use parison control for the production of most PVC bottles and parison programming should only be used in the last resort if other methods fail.

7.3.6 Shaped Dies and Pins

With oval, triangular, square or rectangular section bottles it is often possible to achieve good wall thickness distribution by shaping the pin or die so as to vary the annular gap. A typical example is shown in Fig. 4.

For this rectangular shaped bottle, the die gap is modified to give increased thickness for the corners. Some trial and error shaping may be

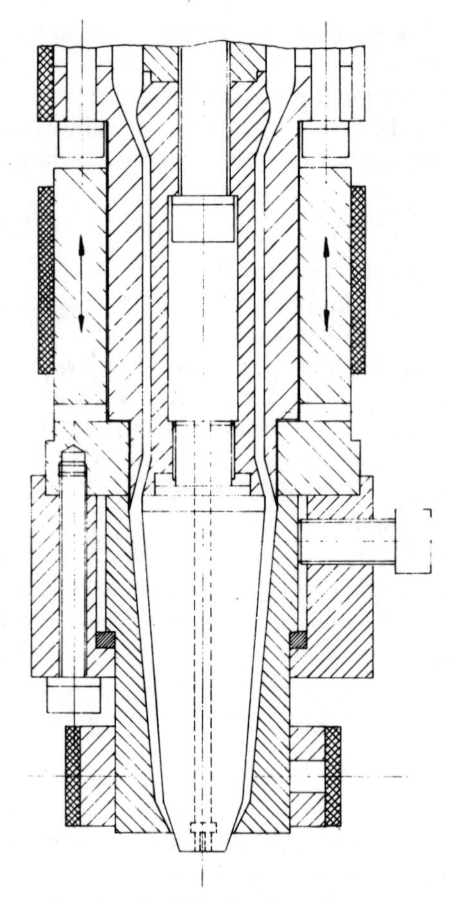

FIG. 3. Variable aperture die.

needed to give the best results, but note that too great a variation in the die gap will create flow problems and blow problems in the final parison.

Shaping the pin instead of the die will yield similar results.

7.3.7 Twin Head Extrusion

Some larger PVC blow moulding machines have been adapted to extrude twin parisons, so as to give higher outputs for the floor space covered. The same machine can often be used with a single head to blow larger bottles. Such a system is very versatile as a wide range of bottles may be produced from one machine.

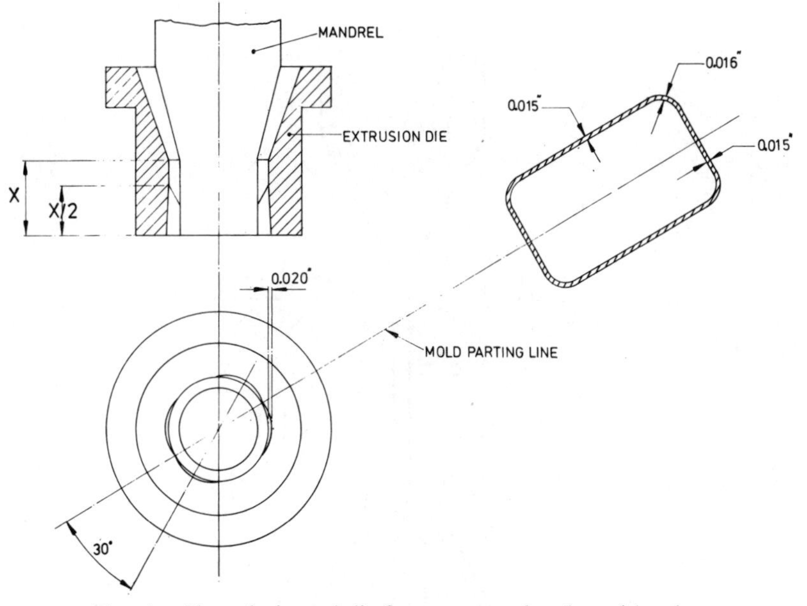

FIG. 4. Shaped pin and die for a rectangular shaped bottle.

As a twin parison machine, it is usually provided with two single heads arranged alongside one another. The separate feeds are taken from a common source which is as near as possible to the material exit from the screw tip. The lengths and resistances of these flow paths must be equal and constant so as to give equal flow paths to the PVC, thus giving equal length parisons.

Accurate temperature control of each head is critical as temperature influences the flow rates. It is not generally possible to use mechanical valving systems to control flow, as is commonly used with polythene, due to the thermal instability of PVC. Variations in temperature, between the heads, will give some control over parison length, but this must be very carefully controlled, and is slow in effectiveness.

Twin extrusion from a single extruder presents other problems of flow which make it more difficult to handle than single head extrusion. For example, if the material leaving the tip of the screw is at a higher temperature than the material nearer the barrel, then the flow pattern through the heads is as shown in Fig. 5. The temperature variation continues right through the heads and when the material emerges from the dies it curls outwards.

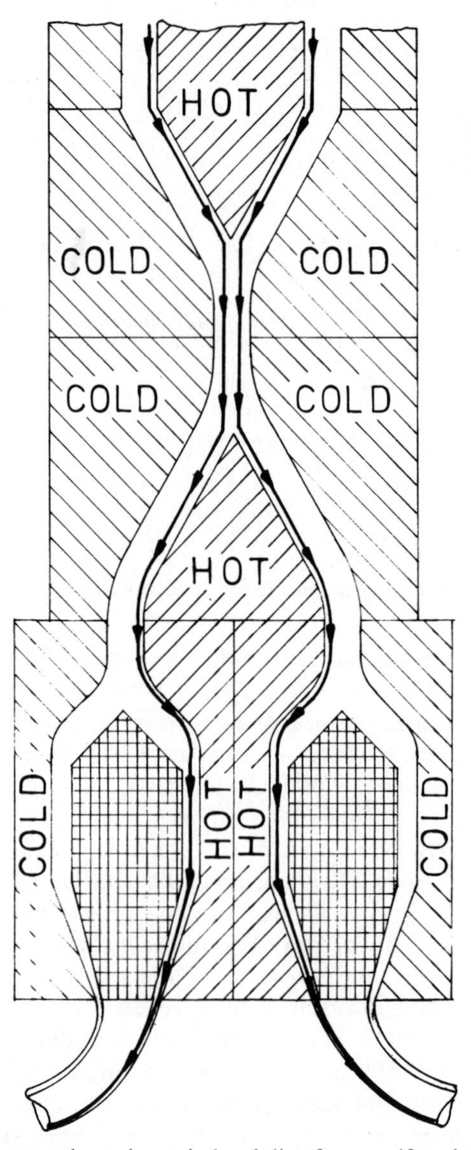

FIG. 5. Flow pattern through a twin head die of non-uniformly heated material.

If the dies are adjusted so as to straighten the parisons then the bottles blown will exhibit a thin section on one side and a thicker section on the opposite side. This can be avoided by cooling the tip of the screw with air, or another cooling medium, until the melt leaving the screw is at a constant and even temperature.

7.4 EXTRUSION BLOW MOULDING UNITS

Various systems to take parisons, clamp them between mould halves and blow them into bottles have been evolved throughout the world and such systems are the subject of numerous patents and copyrights.

Generally, two main systems have developed:

1. Rotary wheel types (carousels) with either vertical or horizontal movements; and
2. Platens which move to and from the head, in a reciprocal manner.

Similarly, blowing of the parisons has evolved into bottom blow, top blow or needle blow techniques, each of which has its own advantages.

In bottom blow systems the parison is usually dropped over the blowpin before the mould closes, and then the bottle is blown. With such a system excess material is trapped between the blowpin and neck portions of the mould giving what is termed a 'flashed neck'. In a top blow system, the parison can be enclosed within the neck diameter before the blowpin is inserted, and this gives a clean or unflashed neck. Needle blown products can be either flashed or unflashed, but in either case a separate trimming operation is needed after blowing to remove excess material and to provide a suitable neck for liquid-tight seals.

Needle blow and bottom blow have the advantage that blowing can take place immediately the moulds are closed. In top blow systems the parison is usually transported to a blow station before the blowpins can enter and blow the bottles. Top blowing does, however, give bottles with a smooth top surface to the neck, and this provides an ideal seal for wadded closures.

It is not my intention to go into detail on the mechanics of the blow moulding unit whether it be mechanical, electrical or hydraulic in operation. However, the mould clamping must be adequate so that the largest size bottle the platens will accept may be blown. Blowing pressures of 10 Bar (145 lbf in^{-2}) may be used. Movements of the units to and from the head must be carried out in the minimum possible time, so that the dry cycle time of the machine is kept low.

Location of the blowing pin is critical, whether it is bottom blow or top blow, and adequate thrust must be available to sever cleanly the waste from the neck portion so as to provide a good cap seal.

Many modern PVC blow moulding machines have been developed in which the top and tail waste is separated from the bottles before they are ejected from the machine. This may sometimes be built into the blow mould tooling itself, or the scrap removal may form part of a secondary process which is built into the machine cycle and which operates before the bottles are ejected. The latest machines present blown bottles in an oriented fashion, standing upright on a conveyor, for inspection, packing or other operations. Even bottles with handles can be deflashed in the machine before ejection.

7.5 BLOW MOULDS FOR PVC

7.5.1 Materials of Mould Construction

When considering blow moulds for PVC, attention has to be given to the acid nature of the gases surrounding hot parisons, and the corrosive effect of these on normal carbon steels. For this reason materials have to be chosen for blow moulds which are corrosion resistant, such as, stainless steels, aluminium alloys, zinc alloys, and beryllium copper alloys.

Consideration must also be given to the potential life required from the tools. The high wear resistance needed by cut-off edges, at the base and neck of the container, must also be appreciated.

Stainless steels will give the highest wear resistance, but the lowest thermal conductivity. Aluminium will give high heat transfer, but is very prone to surface damage and wear. Aluminium and zinc alloys are often used for large moulds where weight is a consideration and casting rather than cutting from solid is being considered. Beryllium copper is often used for inserts in regions where thick sections of a bottle need rapid heat transfer to balance the cooling.

Hardened steel dowels and guide bushes are needed to ensure positive location of the two mould halves together with the required degree of wear resistance. Bushes should have side clearance holes so that any waste PVC which becomes trapped can be ejected automatically. Ease of repair is another important consideration, as a machine breakdown for minor damage to a mould can be expensive if repairs are difficult.

When choosing materials for PVC moulds, a compromise between these factors is usually made, and balanced against an economical cost of manufacture.

7.5.2 Mould Cooling

After shaping bottles in the mould cavity, the second main function of the mould is to remove heat as rapidly and uniformly as possible. Fortunately, PVC cools more quickly than most other thermoplastics, and thin walled containers can be chilled very quickly, thus leading to fast cycle times. However, too rapid cooling of thicker sections can lead to internal stresses which weaken the bottle and result in drop failures.

An advantage of PVC is its low shrinkage, compared with the polyethylenes, and this means that voids and distortion are reduced and size tolerances can be held more accurately.

Cooling channels can either be drilled into the mould block or milled out as a labyrinth cavity, behind the mould, which is then sealed with a back plate. A minimum wall thickness of 8–10 mm (0·32–0·40 in) should be allowed between the mould cavity surface and the cooling channel.

Water flow should not be restricted in any way, and the maximum amount of turbulence which can be induced will assist in heat removal. Flow can be in one continuous channel with one inlet and one outlet, or it can be arranged in three or more separate channels with independent inlets and outlets. If the outlet channels of the latter system are valved, the extraction of heat can be more uniformly controlled. Water at temperatures as low as 5 °C can be used with advantage for cooling PVC bottle moulds.

In addition to the mould there is usually a blowpin which forms the internal diameter of the neck. As this is also in contact with hot PVC it must be cooled equally as fast as the mould. Special designs of blowpin, with water flowing down to the tip and back, have been developed and these give good cooling.

7.5.3 Cavity Venting

When blowing the parison to the shape of the mould, the air trapped in the mould halves must be allowed to escape. If this is not done the bottles will show marks, where pockets of air have been trapped and will have an 'orange peel' surface finish. Trapped air also prevents the PVC from cooling, due to poor contact with the cold mould surface.

In the majority of cases, trapped air can be vented through the mould parting line by a series of slits which are ground to a depth of approximately 0·1 mm (0·004 in). Alternatively, venting channels can be formed, to remove air from various positions, along the split line. In special cases additional vents may be needed for corners, recesses, threads, strengthening ribs, etc., and the need for them is dependent upon the bottle design. Such features should be built in at the mould design stage so as not to interfere with mould cooling channels.

Some PVC materials, particularly calcium zinc stabilised grades, cause wax to be deposited on the cold surfaces of moulds and also on the vented parting lines. This deposit builds up and eventually seals off the designed vents. Solvents must be used on occasions to remove deposits and prevent this happening.

Mould surfaces should generally be smooth and polished as any roughness due to milling marks, sandblasting, etc. will show up on the bottle surface.

7.5.4 Narrow Neck Bottles

Bottles with small necks, relative to the bottle size, can be blown with the neck trimmed and finished in the blowing process. In this case hardened steel plates are fitted above the neck thread, and these cut against the hardened blowpin to trim off the waste. These plates must be accurately sized and hardened to ensure consistent production and they should also be capable of easy replacement if wear occurs.

7.5.5 Wide Neck Containers

Bottles with a wide neck, relative to the bottle diameter, are often blown with a 'lost blowhead'. This must be cut off so as to leave a wide opening in the neck. This was often carried out as a secondary process, away from the blow moulding machine, but the latest machines are designed to hold the bottle rigidly in a fixture (after it leaves the mould) whilst the cutting operation is performed. Bottles leave these machines ready for packing as all waste has been removed.

Firms or companies marketing such systems are Bekum (BAE series), Fischer (FBZ series) and Kautex (KEB series).

An alternative system, patented by Bekum Maschinen-Vertrieb, Mehnert & Co., permits the wide neck to be cut in the mould immediately after blowing, and before the mould opens to release the bottle. A system devised by Sidel uses a separate trimming device set up on a conveyor which takes the bottles directly from the machine and trims off the waste automatically.

7.5.6 Bottles with Handles

Bottles with handles can be blown in PVC provided that the parison is big enough to cover the handle area, thus ensuring that sufficient material is available to form the handle. With through handles the waste portion in the handle has to be removed after blowing. Again, the latest design machines from the major blow moulding machine manufacturers can remove this waste, together with the top and tail waste, automatically in the machine before the bottle is ejected.

Bottles with handles should be designed with the handle well within the body area of the bottle shape so that the required parison size is not too big. The waste portion should be contained within the base area of the bottle if possible.

7.6 INJECTION BLOW MOULDING

An alternative method of manufacturing PVC bottles is by the injection blow moulding process, although PVC does not lend itself too well to this process due to its low thermal stability. A limited number of PVC copolymers and special formulations have however been developed to run on such equipment and some moderate success has been achieved. Special manifolds have been designed to assist the flow of PVC and to limit the heat breakdown problems associated with it.

Basically, the principle of operation of this process is to injection mould a parison shape around a core rod, and whilst the material is maintained in a hot condition, the core rod is transferred to a blow mould where the bottle is blown. When cool, the bottle is ejected from the blow mould and stripped from the core rod. On larger machines a multitude of core rods and moulds increases the output rate of the machine to the limit of its capacity. Core rods are often mounted on a turret which indexes in turn from the injection station to the blow station, and finally to the ejection station (Fig. 6).

Advantages claimed for injection blow moulding systems are:

1. No waste material to be reground;
2. Better neck finish, as the neck is injection moulded;
3. Better wall thickness distribution;
4. Better overall quality and finish; and
5. Less scrap.

Against this is the disadvantage of considerably higher tool costs and the heat stability problems posed by PVC. The latter problem has restricted the full use of custom injection blow moulding machines as the maximum number of cavities which can be moulded on a continuous basis, without the

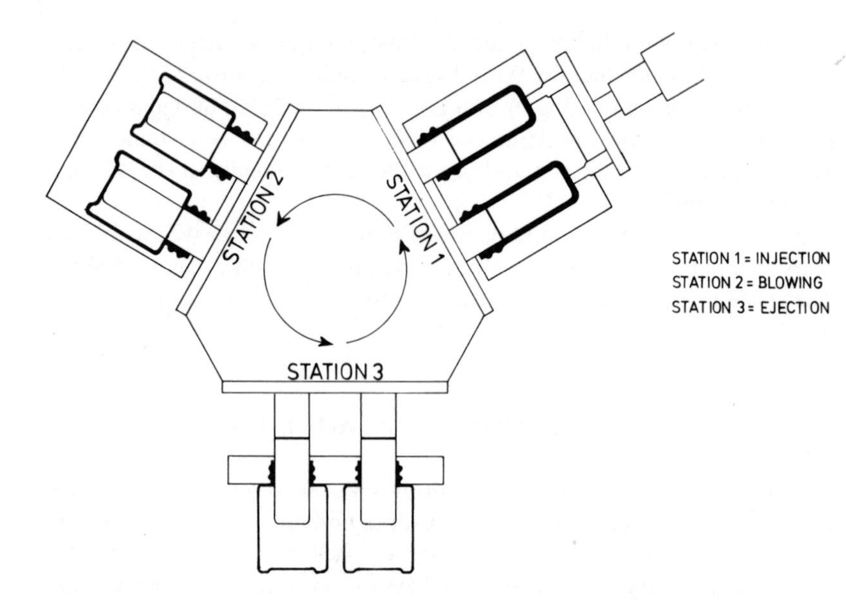

FIG. 6. Turret system used to speed up injection blow moulding.

PVC decomposing, is currently about four. Future developments, in the areas of PVC formulations and manifold design will improve this situation, but until that time extrusion blowing of PVC bottles will remain prominent.

7.7 PVC MATERIALS

7.7.1 Formulations

A wide range of PVC-based materials is available for blow moulding, and an infinite variety of formulations can be chosen so as to suit particular machines, processing conditions, and products.

Popular basic formulations in common use are mostly for clear, unplasticised grades of PVC which are stabilised by organo-tin, thio-tin or calcium zinc stabilisers. Impact modifiers are often added to give varying degrees of impact resistance, with a level at 10% now almost standardised for general purpose use. Lubricants, processing aids and optical brighteners can all be added to give an infinite variety of properties to suit particular applications. The natural slight yellowness of PVC can be corrected by colour tinting in blue, green or neutral shades to enhance the clarity and sparkle for different products. Opaque or translucent colours can be readily produced using the clear formulation as a basis.

PVC blends, used for foodstuffs or pharmaceutical packaging, must meet stringent regulations laid down by the relevant authorities in various countries so as to prevent the use of dangerous substances in contact with food and drink.

Regulations are laid down by bodies such as the:

1. British Plastics Federation (BPF code of practice);
2. British Industrial Biological Research Association (BIBRA);
3. Food and Drugs Administration (FDA); and
4. Bundes gesundheitant (BGA).

As mentioned in Chapter 3, levels of VCM must be very carefully controlled throughout the processes of manufacture, blending, extrusion and blow moulding, so that the level left in the PVC bottle is almost undetectable (current levels in bottles now being manufactured are less than 1 mg/kg (1 ppm)). Manufacturers of PVC polymer have reduced the levels of residual VCM considerably since 1973, and lower levels are forecast for the future thus providing a margin of safety considerably higher than thought possible some years ago.

Food and drink products are also subject to taint and taste, and special formulations have been devised to reduce this to a minimum. Calcium zinc stabilisers are particularly favoured for this use, as they impart the minimum of taint with almost non-detectable smell.

The temperature of extrusion naturally sterilises the inside of bottles, but clean air is often used to remove fumes from inside bottles immediately after blow moulding. Airing of the bottles by storage in clean conditions also assists the removal of smell or taint.

7.7.2 Impact Resistance

PVC is basically a brittle material and bottles blown in unmodified rigid PVC will have a poor resistance to breakage. For this reason it is usual practice to add impact modifiers to enhance these properties and reduce breakage to a minimum.

Levels of impact modifier up to 16 % will improve the drop resistance, as shown in Fig. 7. Levels above 16 % have little practical benefit, and may even worsen the properties of the PVC, by increasing the gas permeability, if taken to too high a level.

Bottle design plays an important part in breakage resistance. Sharp corners, joints or sharp changes in wall section will weaken bottles irrespective of the level of impact modifier used as the material is notch sensitive. Bottles which are round in section are not necessarily the best either and those with a flexible wall which will give on impact are generally better than rigid, cylindrical shapes which do not flex on impact.

Moulding temperatures and processing conditions also affect the impact resistance of bottles, as does of course the quality of the PVC blend in use.

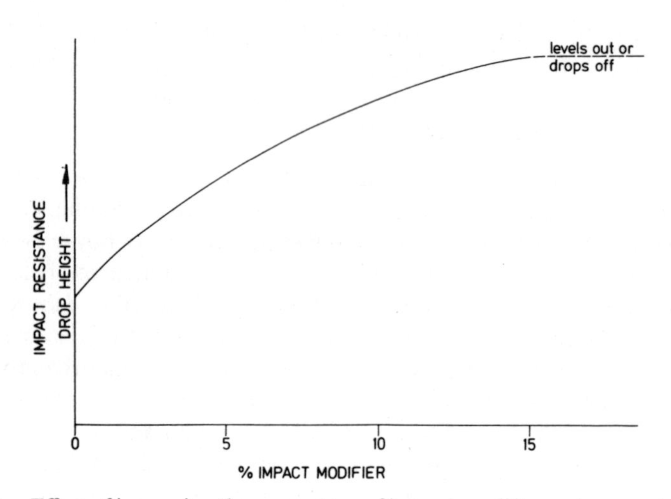

FIG. 7. Effect of increasing the percentage of impact modifier on impact strength.

What is packed within the container also effects the impact strength of the package. For example, bottles filled with liquids perform worse than bottles filled with powder or granular products as these have a self-dampening effect when dropped.

Impact resistance is a selling feature of PVC bottles which should not be ignored, and correct design and use of the right grade of PVC are important factors in creating the right image for plastic containers in the eyes of the general public.

7.7.3 Orientation

Some special techniques have been developed in latter years to increase the strength of PVC by bi-axial orientation. This involves stretching the material, while it is held within a narrow temperature range, in two directions (longitudinally and circumferentially) so as to increase its strength and improve properties such as: stiffness; tensile strength; impact resistance; gas and vapour resistance; burst strength: and transparency. At the same time, the bottle weight may be reduced.

For these reasons bottles formed by this method are suitable for the packaging of carbonated drinks and other products which are pressurised and normally packed in glass bottles or metal cans. Without orientation, the PVC bottle would have to be very heavy, so that it could withstand internal pressure and have the gas barrier properties needed.

Comparative weights of a standard 0·33 litre bottle, which could hold 6 g of carbon dioxide (CO_2)/litre, are quoted in Table 1 by 4P-Verpackungen GmbH who developed the injection blow stretch (IBS) system.

TABLE 1

	Glass	*PVC (Extrusion blown)*	*PVC (Bi-axially oriented)*
Weight (grammes)	165	40	25

It can be seen that, compared with glass, PVC oriented bottles offer big weight savings. There are other advantages, e.g. reduced risk of breakage and less noise on filling lines. Furthermore, it is not necessary to add an impact modifier to the PVC, so that material costs are reduced with a gain in strength and an improvement in gas permeability properties.

Various machines have been developed to manufacture bi-axially oriented PVC bottles, of which the more notable are as follows:

1. *Fischer FIB* 517

This is an injection blow stretch (IBS) machine (Fig. 8) developed and licensed by 4P–Verpackungen, Germany, and manufactured by Johann Fischer, Lohmar Germany. The system is a fully automatic IBS machine with single cavity injection and blow moulds, and has a horizontally rotating table holding 6 core rods.

Temperature conditioning of the parison takes place between the injection and blowing stations. There is no mechanical stretching of the parison, but there is a blow-up and stretch ratio of around 2·4:1. The machine is designed to produce relatively small bottles (up to 0·5 litre capacity) at rates up to 500–600 per hour and as the necks are injection moulded, even thick sections can be of accurate dimensions so that crown cork closures can be fitted even on to the thin-walled bottles.

FIG. 8. Fischer FIB 517 injection blow stretch machine.

The self-contained modular construction of this machine facilitates the gradual introduction of machines so as to build up to the output requirements of any particular plant, thus making it a versatile machine especially suitable for 'in-plant' operations.

2. *Gildermeister Corpoplast System*

The Corpoplast System (Fig. 9) for producing bi-axially oriented PVC bottles is different from the previous system, in that it is a two-stage process rather than a single-stage process. Manufacture of the preforms is done at

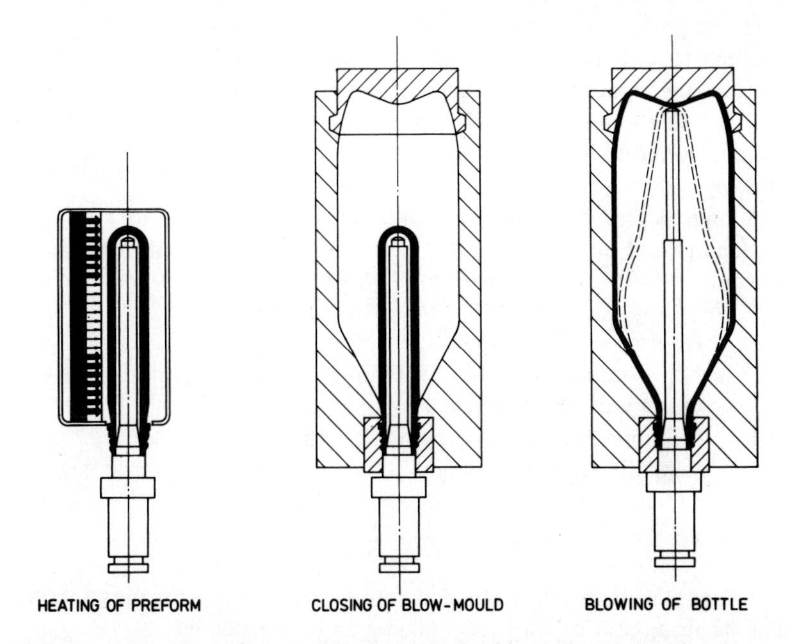

FIG. 9. Corpoplast blow moulding process for bi-axially oriented bottles.

the first stage and this can be performed by either injection moulding or by shaping lengths of extruded tube into moulded preforms. For example preform machine type BAV starts from an extruded tube section and forms one end of the tube into the shape of a hemisphere; the other end is formed with either a threaded neck or a crown closure finish.

The preforms are then allowed to cool and stored in preparation for use on the second stage blow moulding machines.

The second stage machines are of two types, for either:

1. Low volume output (machine type BMB will produce up to 500 bottles per hour); or
2. High volume output (machine type BAB will produce up to 10 000 bottles per hour).

Both these machines will accept either type of preform. Other systems are available which will integrate the preforming and blowing stages at rates up to 10 000 bottles per hour.

Heating of the preforms is carried out by infrared heaters, especially adapted to suit the characteristics of PVC, so that heating is rapid and uniform throughout the thickness of the preform. Inflation is carried out normally with compressed air whilst a telescopic mandrel stretches the material longitudinally and also acts as a central guidance system during the blowing process.

A high degree of bi-axial orientation is achieved due to mechanical stretching and inflation of the parisons at the critical temperature. Advantages of the system are claimed to be as follows:

1. High production rates;
2. No waste re-processing (when injection blow moulding is used);
3. Optimum use of material (bi-axial orientation);
4. Minimum weight/maximum strength containers;
5. Versatile equipment which allows the processing variables to be altered or controlled; and
6. 'In-Plant' production at the filling site.

Against this there are the high investment costs in machines and tooling. The licensing and royalty costs must also be considered.

Machines can be supplied under licence from Gildermeister Corpoplast GMbH, Hamburg.

3. *Bekum BMO Machine*

This is an extrusion blow moulding type of machine (Fig. 10) specifically designed for bi-axial orientation of bottles in PVC and other materials. It has a preform mould in addition to the blow mould, so that the extruded parison can be preformed and conditioned to the right temperature before being transferred to the blowing mould.

FIG. 10. Bekum BMO machine.

The sequence of operations is as follows. The parison is:

1. Extruded and enclosed by a split preform mould;
2. Pinched at the ends;
3. Inflated to the shape of the preform with air;
4. Maintained under pressure within the mould;
5. Transferred to the final blow moulding station at a predetermined temperature;
6. Longitudinally stretched by means of a mandrel incorporated in the blow pin; and
7. Immediately blown and cooled with high air pressure.

Figure 11 illustrates the sequence.

Temperature conditioning is a critical part of the operation. PVC bottles formed by this process have a high degree of bi-axial orientation and strength, and these properties are obtained at significantly lower weights than in the normal extrusion blown bottles.

At the present time this is the only extrusion blow stretch machine, on commercial offer, capable of manufacturing oriented bottles from PVC in a one step process. Outputs vary from approximately 450 bottles per hour

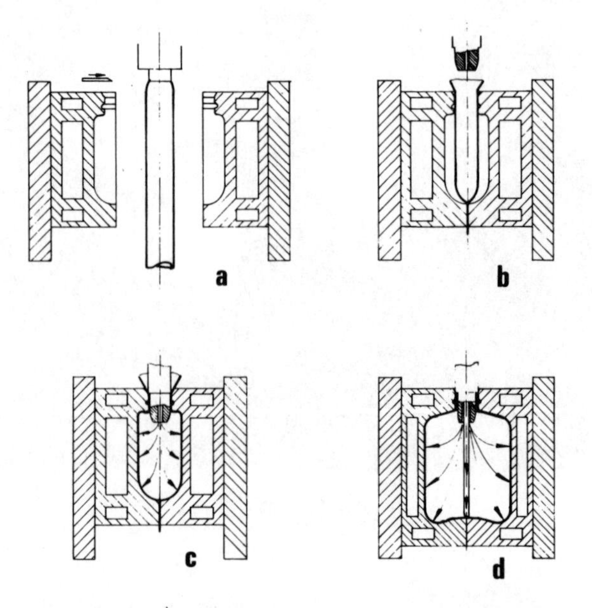

FIG. 11. Bekum blow moulding process for biaxially oriented bottles.

for a single mould set up ($1\frac{1}{2}$ litre bottles), to over 900 per hour for a twin parison machine ($\frac{1}{2}$ to 1 litre sizes). 1800 bottles per hour are possible with a twin parison two-station machine.

Containers leave the machine in a fully finished condition, and the normal Bekum extrusion system for parison programming, top and tail deflashing, etc., can be used to advantage.

Machines are supplied under licence from Bekum Maschinen-Vertrieb, Berlin.

7.8 CONCLUSION

Amongst all the available thermoplastics suitable for blow moulding to-day, PVC stands out as one which offers a unique combination of properties for packaging purposes. It would be difficult to find a substitute which gives equivalent clarity, gloss, impact resistance, gas and moisture vapour resistance, at the economic cost of PVC. In blow moulding, PVC has shown itself to be competitive with other plastic materials. It is also competitive with conventional materials, such as glass and metal, in certain applications.

Despite the difficulties experienced in recent years with VCM, and the carcinogenic associations, PVC has emerged as one of the most chemically 'clean' materials available to-day, for packaging, and continues to be used for food and drink products in most European countries. A large section of the packaging industry is converted to PVC containers, and its usage continues to grow parallel to other thermoplastics.

The future of PVC is therefore well established in blow moulding, and new processing techniques and developments such as bi-axial orientation and co-extrusion with other plastic materials will open up new market areas for future growth.

Chapter 8

THE PRODUCTION OF PVC COATED WIRE AND CABLE

G. P. BARNETT

Francis Shaw & Co. Ltd, Manchester, UK

SUMMARY

The chapter describes the usage of PVC in the manufacture of wire and cable. The extruder and the extrusion process is analysed and its practical performance is derived and applied to the specific case of PVC compounds. The extrusion line for cable manufacture is described showing what components are required and what the limitations are regarding the speeds of production. It is shown that the extruder size has to be compatible with the cable size to be made and the limitations imposed by the line components, and therefore a wide range of extruder sizes both in screw diameter and screw length has to be available from the equipment supplier. It is concluded that cable manufacture is a specialised process demanding very precise control of the extruder and the ancillary equipment, and that automatic control of the process is required to assure good cable quality at high production rates.

8.1 INTRODUCTION

The great traditional materials used for the insulation and protection of electric cable have been rubber and paper and these are still used today although in a decreasing amount. During the war natural rubber supplies were severely restricted and great efforts were made to find alternative materials; impetus was therefore given to the development of synthetic rubbers and especially plastics.

Although polymer chemistry had been recognised as a separate field since the early 1920s, it was the necessity to produce alternatives to rubber that accelerated its progress. Polymer science has given us the vast range of plastics materials that we use today. Unfortunately, many of these uses have been as substitutes rather than as materials having properties which can be exploited in their own right.

It is in the manufacture of cable, where plastics have such excellent properties, that they have become essential materials. Their use has led to the continued development of new and better cable applications. Many types of plastics are used but far ahead of the field are the two thermoplastics—polyvinylchloride (PVC) and polyethylene (PE).

8.2 CABLE MATERIALS

Generalising, it may be said that PE has the better electrical properties but PVC is more versatile and can be compounded to give a wide range of physical properties. PE finds uses in high frequency types of cable (such as telephone and coaxial cable where dielectric loss is of paramount importance) and in high voltage cable (where in addition to dielectric loss, voltage withstand is its major strength).

PVC on the other hand has good electrical insulation properties but these are not good enough for long distance, high frequency work nor for voltages much in excess of 5 kV. It has a major advantage over PE in being self extinguishing, useful where there is a fire risk, and is used within telephone exchanges for switchboard wiring. It finds great use in domestic and industrial wiring cable and as a protective sheath for practically all types of cable.

PE must be used in a relatively 'pure' state to retain its excellent electrical properties but usually an anti-oxidant is compounded into the material to prevent degradation during processing. PE is also degraded by the effect of the ultra-violet radiation present in sunlight, and for outdoor applications a few percent of carbon black is usually included as a filter. The material also suffers from environmental stress cracking, especially in the presence of detergents and solvents, and for this reason the extrusion conditions should be carefully selected and controlled.

PVC does not have these disadvantages to the same degree and in addition has the great advantage that it can be tailor-made for a particular application. The properties can range from rigid to highly flexible and with differing degrees of electrical strength and working temperature range. Table 1 shows a comparison of various cable insulation materials.

A wide range of PVC formulations is available. For example, BS 2746

TABLE 1

THE PROPERTIES OF SEVERAL MATERIALS COMMONLY USED FOR THE INSULATION AND SHEATHING OF CABLE

	A natural rubber compound	*A synthetic rubber compound*	*PE*	*A PVC compound*
Specific gravity	1·5	1·25	0·92	1·33
Volume resistivity (Ohm—m)	10^{13}	10^{9}	10^{16}	10^{10}
Voltage breakdown (kV/mm)	6–14	6–14	21·5	14
	Depending on formulation			
Dielectric constant	3·0	9·0	2·3	7·0
Abrasion resistance	Excellent	Excellent	Good	Good
Flame resistance	Poor	Good	Poor	Good
Flexibility	Excellent	Excellent	Fair	Good
Working temperature range (°C)	−40–70	−30–90	−60–80	−55–105
Weatherability	Poor	Excellent	Good	Excellent

shows several types of PVC and Fig. 1 shows the correlation between these types and their softness or flexibility (expressed in terms of BS softness or Shore hardness).

8.3 THE EXTRUDER

8.3.1 Extrusion Feed

The normal PVC cable compound is produced from the following components:

1. PVC polymer;
2. Plasticisers;
3. Stabilisers;
4. Fillers; and
5. Pigments.

The compounded material is supplied to the extruder in the form of

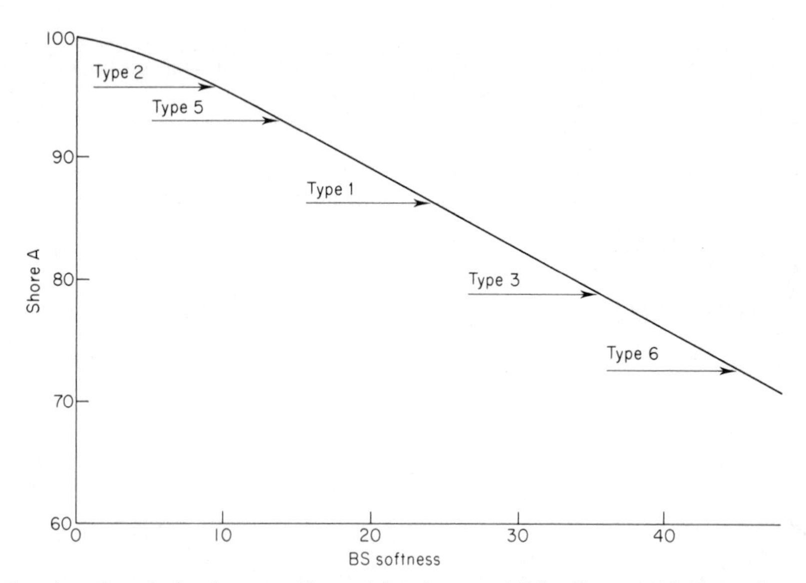

FIG. 1. Correlation between Shore A hardness and BS softness (with the various types of PVC specified in BS 2746 superimposed). Type 1: General purpose insulation; Type 2: Hard insulation for 70 °C operation; Type 3: Transparent flexible insulation; Type 5: Hard insulation for 85 °C operation; Type 6: General purpose sheathing.

granules. A relatively small quantity is used in powder, or premix, form and the mixing process is completed during the extrusion process.

It is generally preferred to use granular feed since this produces a more uniform extrudate. When powder feed is used, special precautions have to be taken to extract both air (which would normally escape via the feed hopper) and volatiles (which would normally escape during the compounding process). Thus either a vacuum hopper or a vented barrel would be used, or even both.

8.3.2 Extrusion Mixing

The extruder is used as a device for softening the PVC granules, mixing them into a homogeneous mass and pumping this melt through a die to form the product. In the special case of cable extrusion the degree of homogenisation is extremely important, in order to obtain a uniform insulation or sheath. Any imperfections are possible sources of either electrical breakdown (in the case of insulation) or mechanical failure (in the case of the sheath).

To achieve this optimum degree of homogenisation or mixing as it is commonly called, special screw designs have been developed to improve on the conventional 'plain' screw. This would have a parallel feed section, a tapering transition section (where the PVC is melted and compacted) and a metering section (where the PVC is mixed and pumped into the head). Such a 'plain' screw would have a channel depth ratio (compression ratio) from feed to metering sections of about 3:1 and would require some back pressure either from the head, or from filters at the end of the barrel, to give adequate mixing. In addition it would usually be necessary to cool the screw by passing water through a hole bored along its axis. This has the unfortunate effect of also seriously reducing the output rate.

Many designs of 'mixing' screws have therefore been invented, by various extruder manufacturers, and these fall into three main classes:

1. Double channel screws in which the melted PVC is allowed to pass from the channel filled with granules into a channel which receives the PVC melt. The volume of each channel is varied along the length of the screw so that at the delivery end of the screw all the PVC has been transferred to the melt channel.
2. Screws with rings of pins located in the metering section of the screw (Fig. 2). These help to homogenise the PVC. The PVC within each channel is normally rolled along by the forward moving flight, giving a circulating motion to the material. This results in well worked PVC near the barrel wall and less well mixed material in the middle of the channel. The effect of the mixing rings is to disturb this regular circulating flow pattern so that all the PVC receives equal treatment. With these screws a long metering section is often used with four or five mixing rings along its length. These continually break up the flow and stop the PVC from settling into a circulating pattern. With this design of screw, water cooling is not required and consequently the output per screw revolution is increased.
3. Screws incorporating smearing sections, through which all the material has to pass after being thermally softened. Thus a degree of uniformity is produced after passage through such a smearing section.

8.3.3 Temperature Control

In all the various types of screw for PVC extrusion the maximum output is limited either by temperature rise or by insufficient mixing as the screw speed is increased. The PVC is, of course, initially fed into a hot barrel and

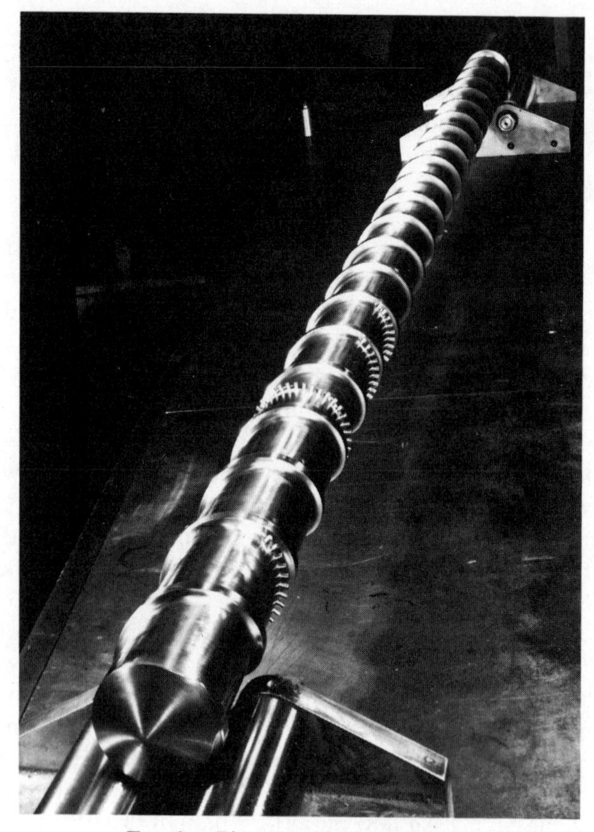

FIG. 2. Pin-type mixing screw.

at slow speeds some heat is transferred from the barrel heaters. However, as the screw speed is increased, heat is generated, due to the high viscosity of PVC, and eventually the problem becomes one of heat extraction. Therefore all barrels are provided with cooling equipment. Cooling is usually achieved by blowing air over a maximised barrel surface area or by passing cold water around the barrel.

Much work has gone into barrel temperature control. Machines are supplied with controllers ranging from simple on/off heating and cooling to sophisticated systems of proportional controllers where the heating or cooling is proportional to the temperature error. A new system has dual thermocouples located deep in the barrel and also in the heater/cooler and measures the heat flow between the two thermocouples. The temperature of the heater/cooler can be predicted and controlled to give an even more accurate temperature control of the barrel.

8.3.4 Theoretical Performance

The theoretical performance of the single screw extruder has been well documented but in practice PVC is very difficult to treat theoretically, and empirical results often dictate screw design and extruder performance. However, some general formulae for extruder outputs are as follows:

$$Q \propto D^2 NhL$$

where Q = Output per unit time
D = Barrel diameter
N = Screw speed (rpm)
h = Flight depth in the metering section
L = Effective barrel length.

The output of the extruder is proportional to the volume of PVC in the final flight of the screw, as $Q \propto h$.

Now in practice,

$$h \propto D^n$$

where n is a constant depending on the viscosity of the material. (In the case of PVC it can vary between 0·6 for flexible PVC to 0·8 for rigid PVC.) Thus, in practice, the maximum output of the extruder is proportional to D^2, and the output per revolution of the extruder is approximately proportional to $D^{(2+n)}$. This is because $Q \propto D^2h$ and $h \propto D^n$. If D^n is put into $Q \propto D^2h$ instead of h, then $Q \propto D^2D^n$. This simplifies to $Q \propto D^{(2+n)}$. It should be noted that the output is proportional to the screw speed and that the relationship is linear. This is especially important for the cable-maker (see Section 8.4).

8.3.5 Length/Diameter Ratios

In practice, barrel length to diameter ratios of 20:1 or 25:1 are commonly used for PVC as these long barrel lengths confer several advantages:

1. The output is less sensitive to back pressure changes (this means that a relatively constant output is obtained over a wide range of cable sizes);
2. The output is greater (approximately in proportion to barrel length);
3. The relationship between screw speed and output is more linear; and
4. Better mixing is obtained.

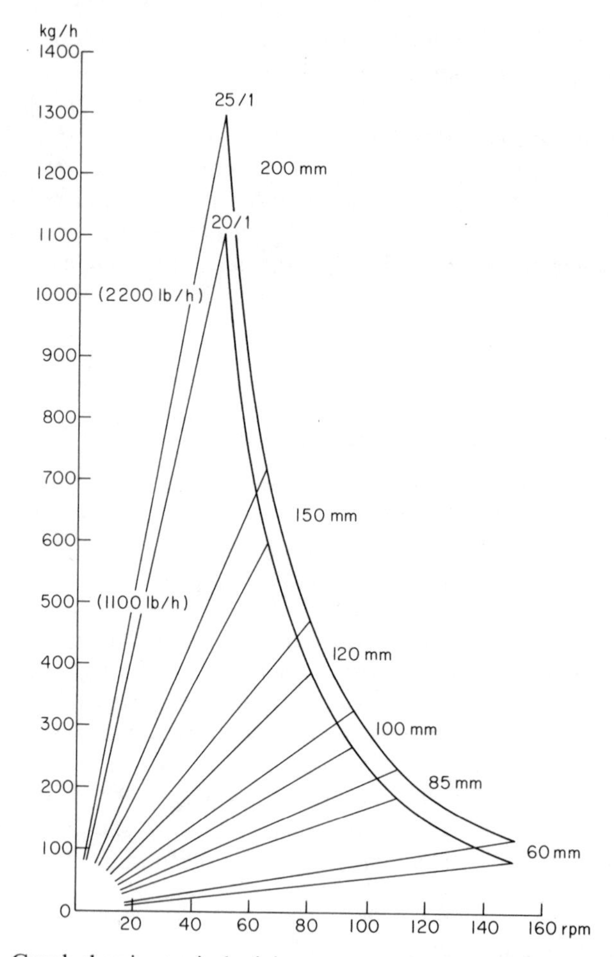

FIG. 3. Graph showing typical mixing screw outputs for different extruder barrel diameters for both 20 L/D and 25 L/D ratios, and the normal envelope of maximum screw speeds.

Typical extruder outputs are shown in Fig. 3. Thus it is possible to choose the size of extruder required for the proposed application, and extruder output, except in the case of large cables, need not be a limitation.

8.3.6 Heads and Dies

The extruder pumps the PVC into a die-head which forms the material round the cable. The design of the head is most important so as to obtain uniformity and concentricity of the insulation or sheath. Head design must be such that there is a uniform flow within the die and to achieve this it is necessary to have uniform pressure across the flow path. In all heads, except the very small ones, streamlined flow is essential to obtain this uniform pressure and also to avoid 'dead spots'. In these spots PVC could be held, decompose and disturb the proper flow path. For such heads it would be always desirable to have a fixed cable guide and a fixed die so as to be always sure of concentricity. However, whilst it is theoretically possible to design such heads, changes in viscosity (due to formulation and output rate differences) make it advisable to provide die adjustment so as to compensate for small flow changes. Thus for larger heads there is usually provision for radial adjustment of the die and axial adjustment of the guide tip. It should also be noted that absolutely concentric tips and dies are not always found, especially after prolonged usage!

On smaller heads, whilst streamlining is still essential, the pressure changes rapidly as the PVC passes through the die. This has an equalising effect so that die adjustment is often dispensed with.

In the case of very small heads streamlining is not essential due to the very high pressures and the very rapid flow of the PVC through the die.

Typical cross-head cross-sections are shown in Figs. 4 and 5.

The arrangement of the tip and die is dictated by the cable structure or formations desired. Two methods are used:

1. Pressure extrusion, when the PVC extrudate is formed between the cable and the die. This produces a tight insulation or sheath which fills any spaces, such as the interstices between conductors (which form a strand) or those between insulated cores (forming a multicored cable, etc.) and still produces a uniform, circular surface.
2. Tubing extrusion, where a tube is formed between the guide tip and the die independently of the conductor or cable. A partial vacuum is produced in the core tube so that atmospheric pressure collapses the tube on to the cable. This method has many advantages and is used wherever possible, for example, for sheathing circular cables and for insulating sector-shaped conductors. Typical tip and die shapes are shown in Fig. 6.

In the case of pressure extrusion, the die diameter should be as close as possible to the required cable diameter. In the case of tubing extrusion the stretching of the PVC from the annulus between the tip and die must be uniform. Therefore, the ratio of the cross-sectional area of this annulus to

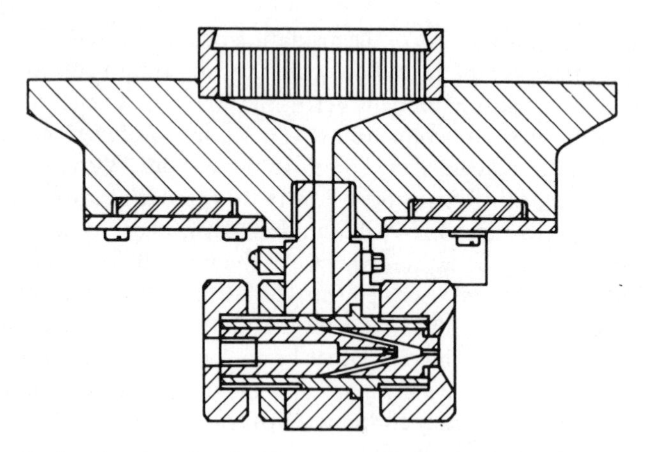

FIG. 4. Outline drawing of typical fine wire cross-head, suitable for insulating fine wire such as telephone wire.

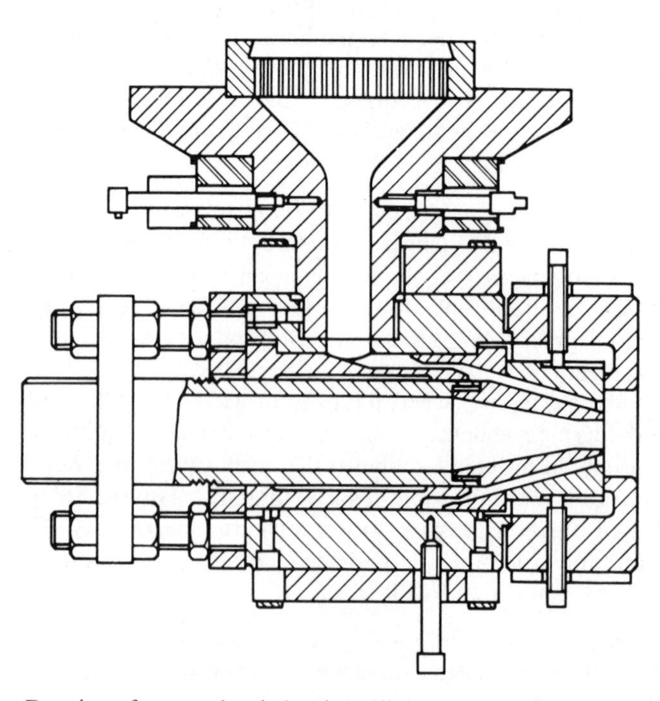

FIG. 5. Drawing of a cross-head showing adjustments and flow paths for cable diameters from 2–150 mm.

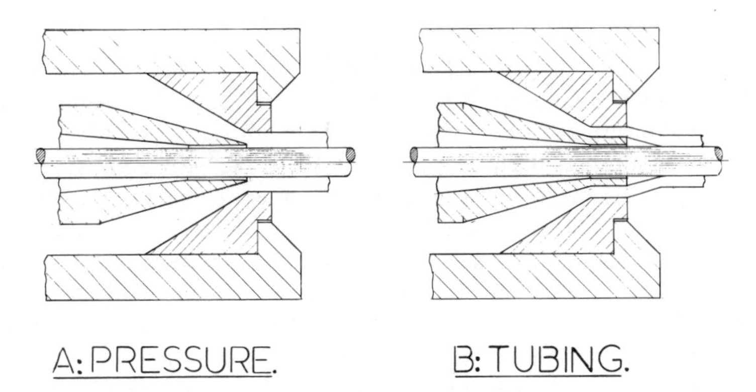

FIG. 6. Drawing showing the relative positions, and typical shapes of tip and die, for (A) pressure extrusion and (B) tubing extrusion.

FIG. 7. The extruder with open head, showing the breaker plate and head clamping arrangement and the hinged head support.

the cross-sectional area of the required PVC sheath must be in the same ratio as the die diameter to the required cable diameter. To reduce the stretch to a minimum, therefore, the thickness of the guide tip should be a minimum consistent with its mechanical strength.

In almost all cases the cross-head is sealed to the barrel, by means of a breaker plate clamped between them, and filters or gauzes are used to remove any contamination present in the PVC. In some cases the filters also serve to increase the barrel back pressure to promote better mixing, but with modern mixing type screws this is rarely necessary (see Fig. 7). The pressure of the PVC in the barrel, which can be up to 750 kg/cm^2 (10 670 lbf/in^2) has to be taken by the head and its clamp. It is therefore necessary to provide some safety device in case there is blockage of the filters or a blockage in the head itself. This may be achieved by using one or more of the following:

1. A shear device in the screw drive;
2. A shear device in the head clamp;
3. Rupture discs;
4. Electrical overload protection.

8.4 CO-ORDINATION OF AN EXTRUSION LINE

The fundamental principles of cable extrusion are deceptively simple:

1. We have to take wire or cable, usually on a reel or sometimes in a coil, and wind it on to another reel or form another coil.
2. As the cable is being thus transferred from one reel to another, we apply an extruded layer of PVC to it and cool the product.

We have already seen that, up to a practical limit, we can obtain the required quantity of PVC by the proper choice of extruder size. Therefore, the optimisation of the extrusion line becomes one of the mechanical problem of pay-off and take-up, transportation and cooling of the cable. It is essential to any cable line that the speed of the cable is as constant as possible since it is the ratio of the extruder output rate to the line speed which gives a cable of uniform size. If the line speed is too slow, the extrudate will be allowed to swell; if too fast, it will be stretched to too small a diameter. Once the correct ratio has been established for any particular cable, we should be able to make the cable at any required speed and still maintain the correct cable size. Modern drive systems can be coordinated such that the speed of the extruder and the line speed hold this constant ratio over a wide range of speeds; hence the requirement of the extruder to give an output proportional to screw speed. The line speed is, of course, controlled by some pulling device independent of the cable take-up, because the latter has to reduce its rotation speed as the reel fills up.

Another desirable feature of a cable extrusion line is continuous operation, which implies joining one length of conductor or cable to another at the input, and transferring the cable from a full reel to an empty one at the take-up. Let us, therefore, now look at the elements of an extrusion line in relation to the function they have to perform.

8.4.1 Pay-offs

Here we have two types:

1. Rotating reel, which can either be driven (at a speed controlled by the tension in the cable) or non-driven (i.e. pulled round by the cable itself).
2. Non-rotating (where the wire is taken off over the flange).

The non-rotating type is used whenever possible since it has the great advantage of enabling the inner end of the reel in use to be brought out and jointed to the other end of a new reel in order to obtain a continuous supply. In practice the two reels are placed side by side in a double cone, which is necessary both to protect the wire and also to contain it. Rotating discs prevent the wire from being damaged by the flange of the reel. This type of pay-off can be used when the twist, which the method imparts to each turn of the wire, can be tolerated. It is, therefore, suitable for small solid conductors, bunched flexible conductors and up to seven wire stranded conductors. The imparted twist would always be in a direction to tighten the existing twist in the conductor. A wire straightener is usually incorporated to even out the twist.

The normal range of reels for flyer pay-offs would be up to about one metre diameter.

In the special case of high speed lines, where a solid conductor is used, the maximum speed attainable, with conventional flyer pay-offs, is about 1500 m/min (4921 ft/min) in practice. To attain higher speeds a wire drawing machine and annealer is placed in tandem with the extruder. This enables the wire to be drawn to the size required at speeds which approach those of the wire drawing and annealing operations when performed separately. This technique has been applied extensively for telephone wire insulation and also for building wire insulation (see Fig. 8).

Rotating reel pay-offs are used for cables which, therefore, cannot be

FIG. 8. A driven disc dual flyer pay-off. In the background is the new reel ready for bringing to the pay-off for jointing.

FIG. 9. A cantilever type pay-off with screw lifting of reels into operating position. The pneumatically controlled disc brake gives tension to the cable and prevents over-run of the reel. These are often used in pairs to obtain continuous operation in conjunction with an accumulator if time is required for jointing.

twisted either by virtue of their construction or their size. Practically all cables for sheathing would come into this category. However, large cables tend to run slowly. This gives time for jointing, if necessary, with the use of an accumulator which stores cable during the time taken for jointing. The common range of reel sizes would be from about 1–3·5 m (3·3 to 11·5 ft) flange diameter and the various types include:

1. Cantilever screw or hydraulic lift;
2. Vertical screw lift; and
3. Portal types where the reel is suspended from above. (See Fig. 9.)

The cable is guided into the extruder cross-head where the PVC is extruded as described earlier, and then the heat has to be extracted.

8.4.2 Cooling

The PVC loses heat both inwards to the conductor or cable and outwards into the water of a cooling trough. The time required for this can be calculated theoretically and very good correlation with practice is obtained. Of course, the cooling time depends on the speed of the cable and the speed depends on the cross-section of PVC compared with the extruder output. However, the cooling length which has to be calculated also depends on the radial thickness (to which the length is directly proportional) and the cable diameter and therefore the surface area (to which the length is inversely proportional). For example, for sheathing a cold cable an approximation for calculating cooling trough length is given by the formula:

$$L = K \times Q \times \frac{R}{D}$$

where L = Trough length (m)
K = Constant (about 0·5)
Q = Extruder output in kg/h
R = Radial thickness PVC (mm)
D = Cable diameter (mm)

(assuming about 25% of the heat remains in the cable core and 75% is extracted by the water). Thus trough lengths of about 20–25 m (65·6 ft–82 ft) are common for sheathing lines based on 120 mm (4·72 in) extruders and about 30–40 m (98–131 ft) for 150 mm (5·9 in) extrusion lines to cool the cable to about 40 °C.

In the case of fine wires, such as telephone wires in the range 0·3–0·9 mm (0·012 in–0·036 in) diameter conductor, the cooling length would be about

35 m (115 ft) at line speeds of up to 2000 m/min (6560 ft/min). Special techniques would be used to avoid water drag and the tunnelling effect of a fully immersed cable. For building wire with conductor sizes in the range of 1·0–16 mm^2 (0·001 5 in^2–0·024 in^2), speeds up to 1500 m/min (4920 ft/min) would be obtained for the smaller sizes quoted, the cooling time is in the range 5–10 seconds; the cooling lengths would thus be about 200 metres (656 ft).

Both the telephone wire sizes and the building wire sizes may be passed through multipass troughs so as to give long cooling lengths in a minimum of floor space (Fig. 10).

FIG. 10. A multipass cooling trough with one or both ends driven so that it also acts as a capstan.

8.4.3 Pulling Devices

These are of two categories:

1. Capstans; and
2. Caterpillars.

Capstans are used for the smaller cables (which can be wrapped round the capstan sheaves) and which are usually made at high speeds. They are usually constructed from dual sheaves, to lift the wire from the main driven sheave, and transport it across the sheave to avoid trapping the turns. An advantage of this type is that the pair of sheaves can be immersed in a water bath thus effectively reducing the trough length. Another type is the belt wrap capstan, which relies on a belt holding the cable against the sheave. In the case of the multipass water trough, one or both ends can be driven so that it acts as a capstan in addition to a cooling trough. For very fine wire special high speed capstans up to 2500 m/min (8200 ft/min) are available (Fig. 11).

FIG. 11. A pair of high speed capstans for metering and tension control of fine wires.

Caterpillars are used for those cables which are too large, or too stiff, to be bent around a capstan. They keep the cable in a straight line. The normal maximum speed of a caterpillar is about 300 m/min (984 ft/min), in keeping with its ability to pull a heavy cable at high tensions. For heavy cables a second caterpillar is often used, between the pay-off and the extruder, to feed the cable at a uniform speed into the extruder head. The second caterpillar, which follows the cooling trough, exerts a constant tension to the cable, thus giving a uniform speed and tension whilst the cable passes through the cross-head.

FIG. 12. A caterpillar for larger cables with a strengthened rubber traction belt supported by articulated multi-roller shoes to give uniform pressure and traction to the cable.

Generally, the longer the caterpillar the higher is the maximum tension. There are some applications where a low pressure on the cable is required to prevent either damage or deformation, and therefore a longer caterpillar can give the required pull at a lower pressure (Fig. 12).

8.4.4 Take-ups

The design of the take-up is largely determined by the size of the reel to be used. Also, large cables are usually wound on to large reels and small cables on to small reels. Since small cables can also be handled at higher speeds than large cables, we have a self-compensating effect. For example, fine wires such as the telephone wire sizes are usually wound on to reels of 400 mm (15·7 in), 560 mm (22 in) or 630 mm (24·8 in) flange diameter. Take-ups are available which can handle them at speeds up to 2500 m/min (8200 ft/min) and which are fully automatic in operation (Fig. 13). Feeding in empty reels, changing from full to empty reel and delivering the full reel from the machine are included. For such sizes the output given by a 60 mm (2·36 in) or 85 mm (3·35 in) extruder would be appropriate.

FIG. 13. Automatic dual reel take-up normally used for telephone wire in the range 400–630 mm reel diameter, capable of speeds up to 2500 m/min.

FIG. 14. Automatic dual reel take-up normally used for building wire cable sizes for reels in the range 600–1500 mm diameter, at speeds up to 1500 m/min.

For building wire sizes, the cable is usually wound on to reels of 630 mm (24·8 in), 710 mm (28 in), 1000 mm (39·4 in) or 1250 mm (49·2 in) flange diameter and speeds up to about 1500 m/min (4920 ft/min) are available with automatic reel changing machines. For these cables, a 120 mm (4·7 in) or 150 mm (5·9 in) extruder would be appropriate. There are, of course, many lines in operation running at much slower speeds where simplicity of operation and lower capital costs are required (Fig. 14).

Sheathing of small cables, for example building wire and domestic flexible cable, can be carried out at relatively high speeds on semi-automatic take-ups. Larger cables are usually wound on to reels up to 2000 mm (78·7 in) flange diameter and dual reel take-ups would be used with semi-automatic operation. Accumulators are sometimes used when it is necessary to actually stop the reels between changeover, e.g. to thread the cable through the flange for testing purposes when winding directly on to a despatch reel.

Larger cables are usually sheathed and wound up on to despatch reels on single reel take-ups, and sizes are often in the range 2 m (6·56 ft) to 3 m

FIG. 15. **Single reel take-up for reels in the range 1000–3000 mm diameter** normally used for sheathing larger cables. These are often used in pairs to obtain continuous operation with manual change-over of the cable from the full reel to the empty reel.

(9·84 ft) flange diameter. A commonly used working rule is that the barrel diameter of the reel should be at least 20 times the cable diameter. Usually the barrel diameter is about half the flange diameter. Of course, for special cable constructions there are exceptions and then special reels are used. Figure 15 shows a single reel take-up for large cables. A large sheathing line would often be limited, in speed, by the extruder output. Appropriate extruder sizes would often be 150 mm (5·9 in) or 200 mm (7·9 in) diameter.

All take-ups need speed control, since the line speed is constant and controlled by the capstan or caterpillar, and the reel must run slower as it fills up layer by layer. Small wires and cables, run at high speed, usually have an accumulator which imparts a constant tension to the cable and at the same time operates, for example, a potentiometer when the speed of the take-up has to be adjusted. Larger cables, on larger take-ups at slower speeds, can be controlled by a simple catenary control arm, which gives a sufficiently long response time for adequate control. When the line speed is too fast for a catenary arm and the cable is too large or stiff for a convenient multi-sheave accumulator, the take-up drive is arranged to operate to give a constant tension in the cable.

8.5 CONCLUSION

Cable manufacture is an exercise in control and the control of an extrusion line can be made automatic by the use of appropriate instrumentation. It is possible to make this instrumentation such that once the various parameters have been set by the operator, any error will be self-correcting.

Thus the operator would be required to set appropriate values for the following:

1. Extruder temperatures;
2. Extruder screw speed;
3. Line speed;
4. Wire preheat temperature (where used);
5. Cable diameter;
6. Cable capacitance (when used);
7. Cable concentricity;
8. Test voltage; and
9. Cable tension.

He would also control the replenishment of the extruder hopper with the correct colour and type of compound or masterbatch.

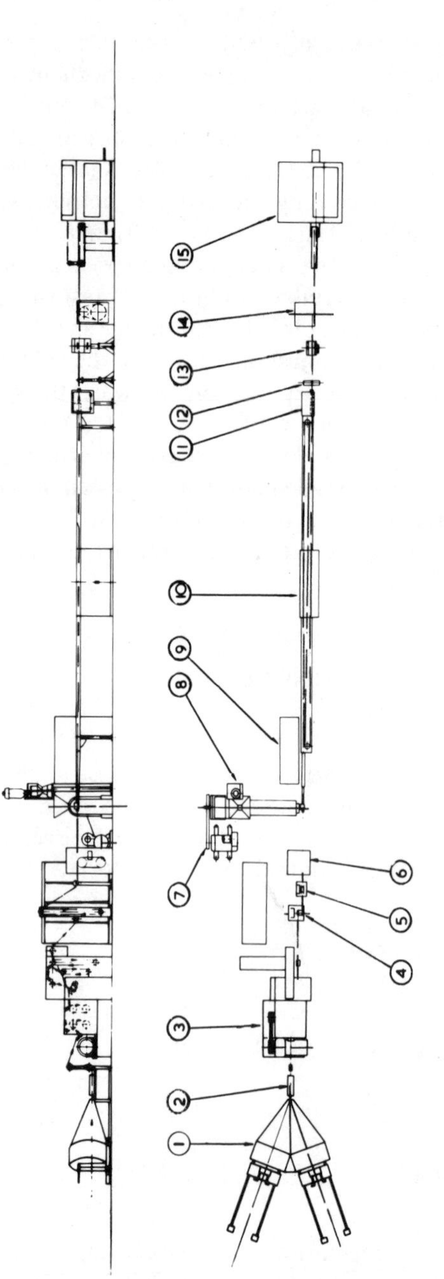

Fig. 16. A typical high speed telephone wire insulation line. *Key*: 1—dual reel flyer pay-off, 2—wire straightener, 3—wire drawing and annealer, 4—accumulator to control wire drawing speed, 5—guide sheave, 6—wire preheater, 7—extruder, 8—hopper loader and masterbatch feeder, 9—control panel, 10—multi-pass cooling trough, 11—capacitance controller, 12—diameter controller, 13—spark tester, 14—high speed capstan, and 15—automatic dual reel take-up.

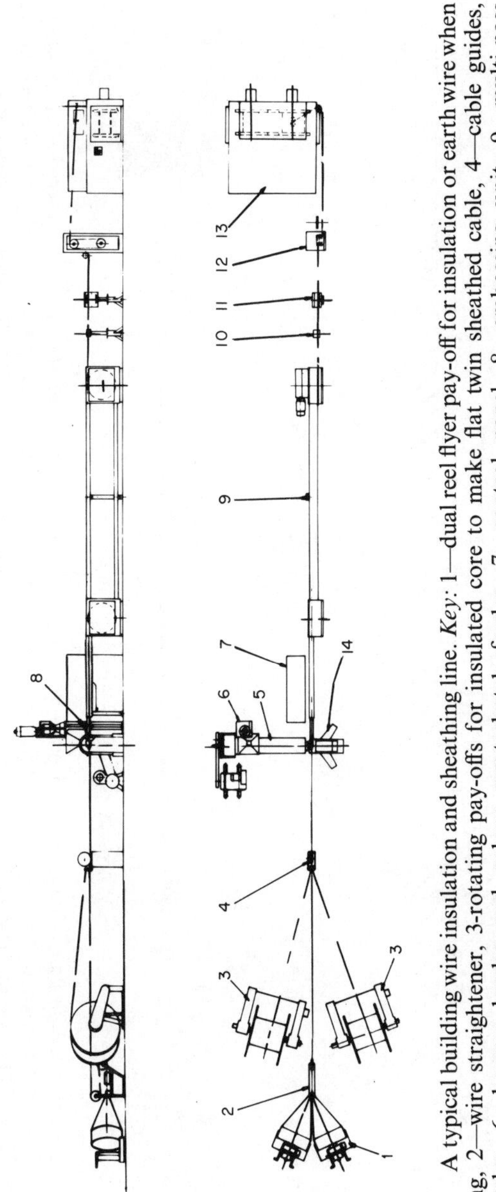

FIG. 17. A typical building wire insulation and sheathing line. *Key*: 1—dual reel flyer pay-off for insulation or earth wire when sheathing, 2—wire straightener, 3-rotating pay-offs for insulated core to make flat twin sheathed cable, 4—cable guides, 5—extruder, 6—hopper loader and colour masterbatch feeder, 7—control panel, 8—embossing unit, 9—multi-pass capstan/cooling trough, 10—diameter control, 11—spark tester, 12—accumulator to control take-up speed, 13—high speed dual reel take-up, and 14—striping extruder.

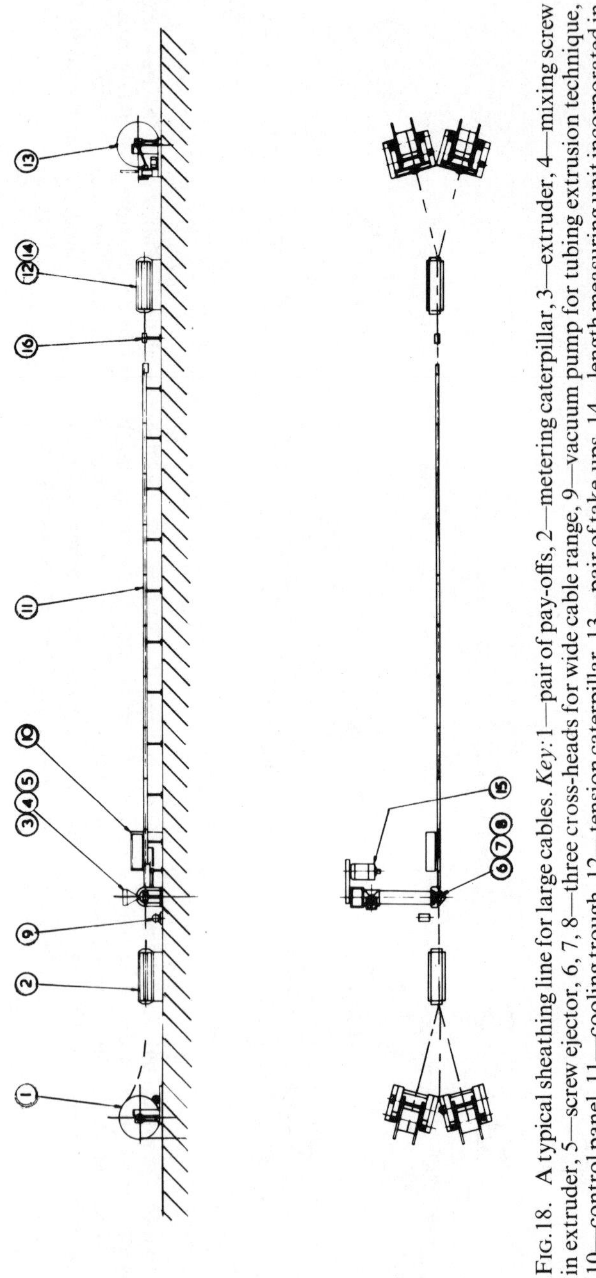

FIG. 18. A typical sheathing line for large cables. *Key*: 1—pair of pay-offs, 2—metering caterpillar, 3—extruder, 4—mixing screw in extruder, 5—screw ejector, 6, 7, 8—three cross-heads for wide cable range, 9—vacuum pump for tubing extrusion technique, 10—control panel, 11—cooling trough, 12—tension caterpillar, 13—pair of take-ups, 14—length measuring unit incorporated in the caterpillar, 15—D.C. drive, and 16—spark tester.

It is also possible to perform the setting of all these parameters and more besides, by the use, for example, of punched cards to make the line completely automatic with virtually nothing left to the discretion of an individual operator. Theoretically, this would make it impossible to produce reject cable. In practice, this degree of automation has been applied to specialised lines, such as telephone insulation lines, where the number of different types and sizes of cable is limited and the quality standards especially demanding.

Thus we have a wide choice of the elements of a cable extrusion line and these may be put together in a large number of combinations, depending on the various factors discussed.

Examples of the various types of line are illustrated in Figs. 16–18.

Chapter 9

THE USE AND MANUFACTURE OF UNPLASTICISED PVC PIPES

J. B. PRESS† and D. A. TREBUCQ‡

† *Wavin Plastics Ltd, Ashford, Kent, UK*

‡ *Wavin Plastics Ltd, Hayes, Middlesex, UK*

SUMMARY

The most common polymer used in the United Kingdom in the manufacture of pipe is polyvinylchloride (PVC). This is used in particular for making pressure pipe for water supply from sizes $\frac{3}{8}$ in diameter to 24 in diameter. In addition, PVC is the standard material used for plastics plumbing products and rainwater goods, with the exception of waste connections between bath, sink or washing-machine and the soil stack. Furthermore, other applications of PVC in pipe-making include ducting and conduiting for telephone and power supplies, perforated or slotted pipe for land drainage and to a lesser extent headers for irrigation purposes.

This chapter will survey the extent of usage and application of PVC pipe and will indicate the tests performed on such pipes. The manufacture of PVC pipes will then be discussed.

9.1 INTRODUCTION

The development of the use of unplasticised PVC (uPVC) piping for the conveyance of cold water and for process line (in pressure) applications, and for rainwater goods, soil systems, underground drainage and for sewers in non-pressure applications has been impressive. From some 3000–4000 tonnes in 1960, the market expanded rapidly until it attained some 100 000 tonnes in 1973 (Table 1). With the current economic recession, production dropped back to 83 000 tonnes in 1975 but preliminary figures would

TABLE 1

TOTAL uPVC PRODUCTION OF PIPES AND FITTINGS IN TONNES, AND PERCENTAGE OF OUTPUT FOR EACH APPLICATION

	1960	1965	1970	1973	1975	1976
Total production	5 000	27 000	67 000	100 000	83 000	95 000
Percentage of output						
Pressure pipe	—	—	50	41	29	—
Rainwater goods	—	—	18	17	18	—
Soil systems	—	—	15	19	20	—
Ducts and conduit	—	—	10	10	14	—
Drainage	—	—	7	13	19	—

Percentages are not available for 1960, 1965 and 1976.

indicate that in 1976 a good recovery had been made with production running at about 95 000 tonnes.

Pressure pipes still represent the main tonnage market despite a considerable reduction not only of output but also of their share of the total production, between 1973 and 1975, from 41 000 to 24 000 tonnes. Production of drainage pipe has risen steadily during this period, probably due to the introduction of uPVC pipes for land drainage applications. The production of rainwater goods and soil systems has remained reasonably constant, no doubt due to the high market penetration: 90 % for rainwater goods, 75 % for soil systems.

It should also be remembered that the specific gravity of uPVC is low in comparison with, for example, cast iron (1·4 for uPVC against 7·2 for cast iron). As an approximation, therefore, the quantity of uPVC pipe and fittings produced in 1973 would equate to some 500 000–600 000 tonnes of cast iron. It would be fair to state therefore that uPVC pipes have now become a well established product making a useful contribution to society.

9.2 DEVELOPMENT OF uPVC PIPES FOR PRESSURE APPLICATIONS

9.2.1 Jointing

In the light of the successful experience in the use of uPVC pipes on the Continent, manufacture in the United Kingdom first started in the early 1950s. Because of its high strength and excellent chemical resistance[1] it found a ready market in the chemical industry, where it showed considerable economies by replacing stainless steel pipes in many applications. The available range of pipe sizes extended up to 6 in (150 mm) diameter and jointing was effected by means of socket and spigot joints, the socket being thermoformed to close tolerances and the assembly fused by means of a solvent cement. These solvent cements were generally based on methylene chloride into which PVC polymer was dissolved.

This system of jointing has proved entirely successful over the years, but as with any chemical joint, it requires a degree of skill and it becomes progressively more difficult to achieve a satisfactory joint with increase in diameter. However, pipelines up to 24 in (600 mm) diameter have been assembled using this technique and are operating satisfactorily.

Fittings were fabricated either by thermoforming or welding or by a combination of both. Welded structures were not, however, very suitable as the weld area tended to be brittle and considerable thought had to be given to the design and installation of the pipeline to prevent such components being subjected to bending stresses. The introduction of injection moulded fittings which could be solvent-cemented to the pipe was a major technical breakthrough in ensuring the integrity of the pipeline.

With increase in production and the consequential reduction in prices, it was possible to envisage that uPVC pipes might compete with pipes made of cast iron and asbestos cement. This was of great interest to the water industry, since it was unlikely that PVC pipes would suffer from chemical attack from acidic soils, etc., which would therefore considerably reduce their maintenance costs. Between 1957 and 1959, a number of experimental lines of 2, 3, 4 and 6 in (50, 75, 100 and 150 mm) diameter[2] were installed. These pipes were assembled by means of solvent-welded socket and spigot joints and after 20 years in use, are still operating satisfactorily. In 1962, BS 3505: 'PVC Pipes for Cold Water Supply', was first published, and has since then had two major amendments issued, in 1964 and 1972, and a complete revision in 1968.

9.2.2 Creep

In considering the hydraulic properties of uPVC pipes, note must be taken of its creep characteristics. The pipe dimensions are calculated from what is now known as the ISO formula

$$P = \frac{2\sigma e}{D - e}$$

where P = 50 years safe maximum hydraulic working pressure at 20 °C,

$\sigma = 50$ years safe maximum working stress at 20 °C, $D =$ outside diameter of pipe, and $e =$ wall thickness.

It can be seen from the above that it is essential to establish a safe value for the working stress (σ) if the pipe is to have a useful life.

It was recognised in the early days of the development of uPVC pipes for pressure applications that, whilst the material could withstand high instantaneous burst pressures, its burst strength tended to decrease with increase in time. Studies were therefore undertaken on sample test pieces and it was noted that, under constant load, deformation would occur and would continue to develop as long as the load was applied, until failure occurred. This phenomenon is known as creep and typical creep curves at different loads are illustrated in Fig. 1.

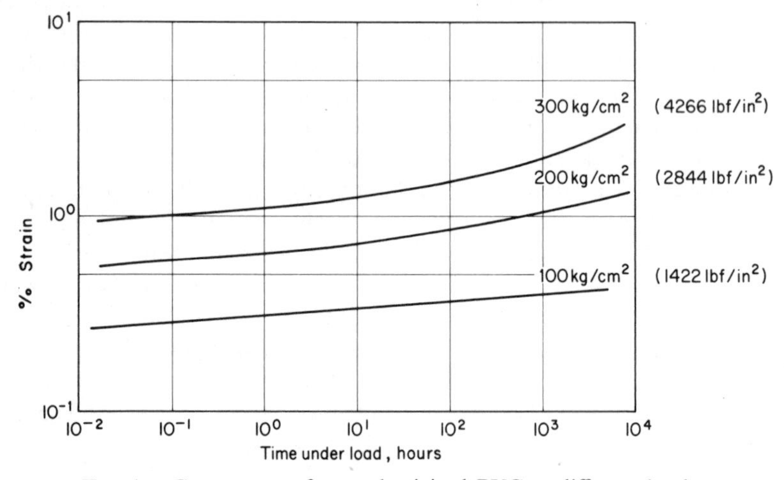

FIG. 1. Creep curves for unplasticised PVC at different loads.

It is obvious that, in the light of this characteristic, information is required on the relationship between the burst pressure of a pipe and its time to failure. Considerable work has been carried out on this subject and has been reported upon in the 1950s and 1960s.[3,4,5] Typical data obtained on pipe manufactured today are illustrated in Fig. 2.

Having established the relationship between pressure and time to failure, the 50 year burst pressure (or burst stress) can be deduced by extrapolation of the data obtained. Whilst this may be a dangerous assumption to make on limited data, tests extending over 100 000 hours (11 years) have been recorded and would support the contention that such regression lines are

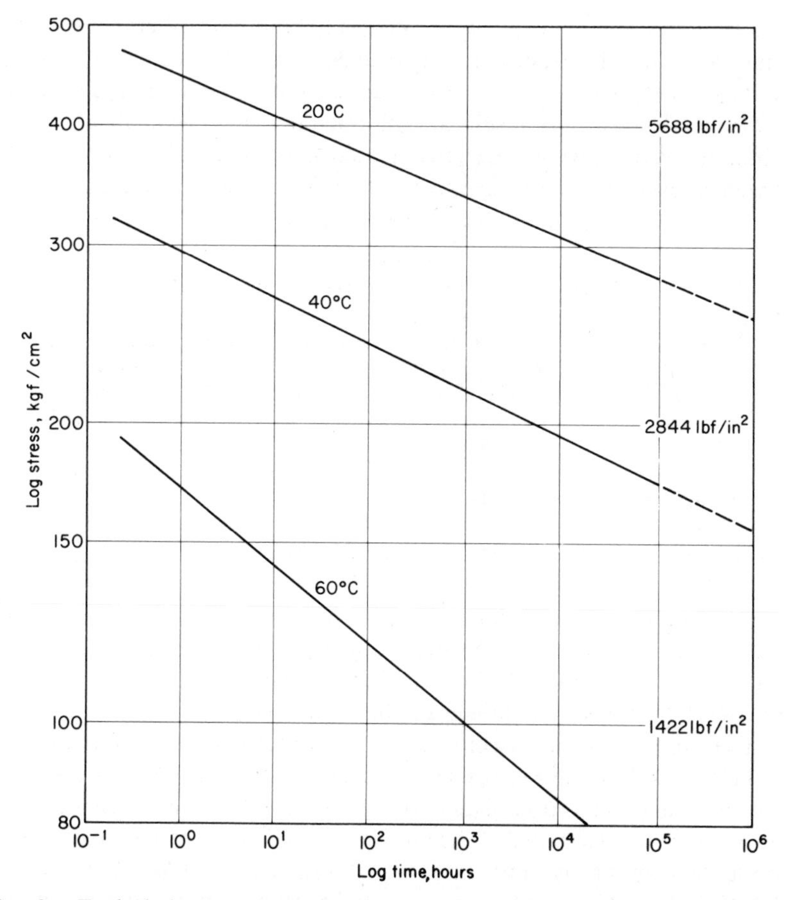

FIG. 2. Typical burst stress diagram for unplasticised PVC at different temperatures.

linear for the higher molecular weight extrusion grades of polymer. Knowing the probable 50 year burst stress of the material, a safety factor can be applied to obtain the safe working stress, a value which can be inserted in the formula referred to above.

9.2.3 Testing

The current edition of BS 3505, the standard for unplasticised PVC pipes for cold water supply, requires pipe samples to be regularly subjected to a long term hydrostatic test. This test requires pipes to be tested at such pressures that some will fail between 1 and 10 hours and others between 100 and 1000 hours. This test is carried out at 20 °C. Having calculated the stress and recorded the time to failure, the data are plotted on a log (stress) versus log (time) diagram. The results are subjected to regression analysis so as to obtain representative extrapolated 1 hour and 50 year burst stress levels which should be equal to or greater than the values in Table 2.

TABLE 2
LONG TERM HYDROSTATIC TEST REQUIREMENTS

Nominal size of pipe (in)	*Minimum 1 h burst stress (bars)*	*Minimum 50 year burst stress (bars)*	*Safe working stress bars (bars)*
Up to $\frac{3}{4}$	353	206	98
1 to 7	396	230	110
8 and over	443	260	123

1 bar = 10^5 N m^{-2} which is approximately 14·5 psi.

It can be seen, from columns 3 and 4, that the value of the 50 year burst stress is more than twice the value of the safe working stress at 20 °C. As indicated in the standard, the use of three different stresses is to provide greater robustness in handling for the smaller sizes.

These data apply only to pipes conveying water at a temperature of 20 °C or below. As shown in Fig. 2, increase in temperature will decrease the stress which the material can withstand for a given period of time. Chemicals may also reduce the resistance of the pipe to internal pressure and guidance on both these points is given in *Code of Practice for Plastics Pipework*, CP 312.[1] The effect of surge pressures must also be taken into account and attention is drawn to the very interesting work of Gotham and Hitch[6] on this subject whose conclusions can be recapitulated as follows:

The results suggest that if the peak working pressure is considered, the design stress in BS 3505 (123 bar) is safe for pipe which is subjected either to a continuous internal pressure or to one in which the pressure varies between ±25 % of the continuous pressure. However for pipe subjected to cyclic loads of high frequency and greater amplitude, it would seem prudent to reduce the design stress perhaps to 60 bars.

Some concern, however, has been expressed that pipe failures can occur at comparatively low stresses over varying periods of time. This failure mechanism has been identified in other materials which are invariably of a brittle nature. This is different from the pattern of the failure mechanism of the creep rupture analysis at 20 °C described above, which is mainly ductile. It is interesting to note that considerable emphasis is placed in the International Standards Organisation's Technical Committee publication ISO/TC138, which deals with thermoplastics pipes and fittings, on testing at elevated temperature. In the case of uPVC this is done at 60 °C. In the authors' experience most of the failures in this test tend to be of a brittle nature and such a test might well be worth studying as a measure of the brittle strength of the material.

9.2.4 Size Range Extension

Over the years the size range has extended. In 1968, the British Standard was extended to cover diameters up to 24 in (600 mm). However the successful use of uPVC pipes would not have been possible were it not for the introduction of joints in which the seal was effected by an elastomeric ring housed in the groove of the socket. A range of these joints was introduced during the middle 1960s. All that has to be done to make an effective joint is to insert the pipe spigot into the socket thus compressing the rubber ring between the pipe spigot and the socket inside diameter.

9.2.5 uPVC for Gas

During the middle 1950s consideration was also given to the possible use of uPVC for the conveyance of coal gas. Pressures were generally low and would not have presented a problem. However coal gas contains various by-products, and additives such as odorants, which were known to attack PVC. The question was whether, in the concentrations in which these chemicals were present, they would weaken the pipe, resulting eventually in failure. Initial investigations proved satisfactory and quantities of pipe in the 1 to 6 in (25 to 150 mm) range have been installed. Later it was decided that a higher integrity could be given to the product by incorporating an impact modifier into the PVC compound, so as to give the pipe better impact strength (approximately an eight fold increase). These modifying agents decreased the burst pressure characteristics of the product (Fig. 3) but the margin of safety was such that this could be ignored.

Subsequent investigation, however, showed that under certain conditions of strain and at certain concentrations of these aggressive products (which were mainly aromatic hydrocarbons) crazing of the inside surface of the pipe could occur. Intensive study of the effect of these crazes on the properties of the pipe indicated that only impact strength and elongation at break were reduced to any marked extent. This phenomenon is referred to

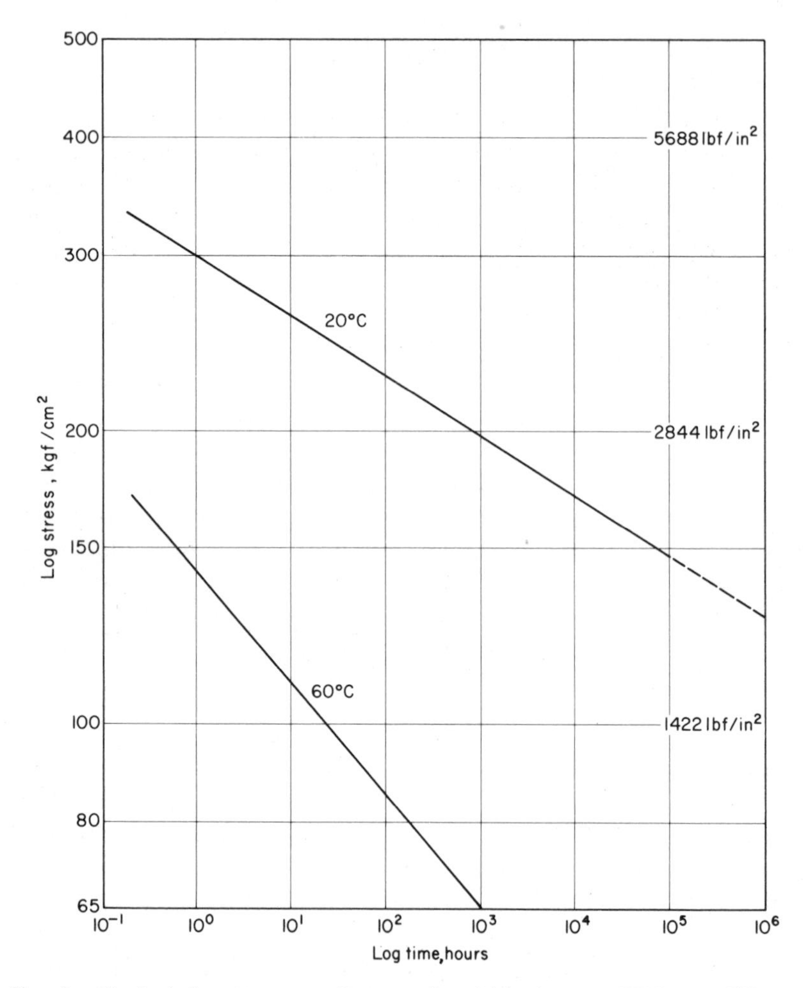

FIG. 3. Typical burst stress diagram for high impact PVC at different temperatures.

as environmental stress cracking and whilst a few failures from this cause have been reported on the European Continent, the risk of failure appears to be very low. Nevertheless, no doubt in the interest of safety, the use of PVC piping for the conveyance of gas has dropped to zero in the United Kingdom. Medium and high density polyethylene, of which there has been many years' experience, particularly in the USA, has taken its place.

9.3 DEVELOPMENT OF uPVC FOR NON-PRESSURE APPLICATIONS

When new materials are introduced to replace well established products, methods of test have to be developed to demonstrate convincingly that such materials are suitable for this particular application. The way this was done for pressure applications has been explained in Section 9.2.2. For non-pressure applications, such tests can sometimes be rather complex and the aim of this Section is to try to explain how some of these tests were evolved.

Figure 4 shows a typical layout of the pipework to which reference is made in this Section.

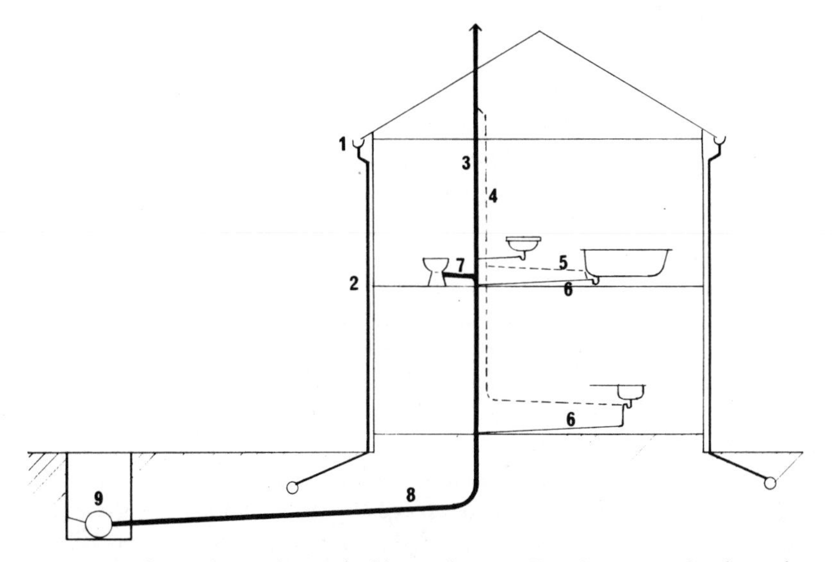

FIG. 4. Sanitary pipework—typical house layout. *Key:* 1—gutter, 2—downpipe or rainwater pipe, 3—main discharge or soil pipe, 4—vent pipe, 5—branch vent pipe, 6—branch discharge or waste pipe, 7—branch discharge pipe, 8—drain, 9—public sewer.

9.3.1 PVC for Rainwater Goods

PVC gutters and downpipes were introduced into this country in the late 1950s and have proved entirely successful. In the early days some systems caused problems because the gutter joint was not always designed correctly and also because the installer did not always appreciate the problems that the high coefficient of expansion of the material ($5–6 \times 10^{-5}$ per degree centigrade rise in temperature) could cause. These problems were, however, quickly overcome. Downpipes were never a problem; their requirements are not very onerous and joints were made by some form of dry interference socket and spigot assembly.

In considering the requirements that a gutter should be able to withstand, it was decided that there were three basic points:

1. The gutter should be able to withstand the abuse created by a ladder being placed against it;
2. When correctly supported, the gutter should be able to withstand the weight of snow from the roof; and
3. Under conditions of expansion and contraction the joints must remain watertight.

Three tests were developed to check for these points and are fully described in BS 4576.[7]

The first test, 'resistance of gutter assembly to impact', requires that a gutter system consisting of a gutter, a gutter joint and joint bracket shall be fixed to the upper surface of a baseboard so that the outer edge of the gutter is offered upwards. A flat headed striker weighing 1·25 kg (2·75 lb) is dropped from a height of 1 m (3·3 ft) on to the outer edge and after impact the assembly shall remain intact and shall show no permanent damage or deformation such as to render it unsuitable for its purpose.

The second test, 'resistance of gutter assembly to load', requires two lengths, each of approximately $1\frac{3}{4}$ m (5·74 ft) to be assembled and correctly supported. A mass of sand, the quantity varying according to gutter size but ranging from 122·5 kg (270 lb) for a 4 in (100 mm) gutter to 171·5 kg (377 lb) for a 6 in (150 mm) gutter, is placed in the gutter and after 14 days the deflection at any point should not exceed 19 mm (0·75 in). The sand is then removed and after 48 hours the deflection, in relation to its original position, should not exceed 3 mm (0·12 in). Finally the gutter is filled with water and allowed to stand for 1 hour. At the end of this period none of the joints shall have leaked.

The third test, 'watertightness of gutter joints', requires two lengths of gutter to be assembled in the normal way with a stop end at each end, fixed firmly so that any expansion which may take place is directed towards the centre joint. The temperature of the surface of the gutter is then raised to 60 °C and maintained at this temperature for 5 minutes. The gutter is then filled to its seal level, with water at 5 °C, and left for 5 minutes. This cycle is repeated 10 times and the joint must remain watertight during the test.

Obviously other tests are included in the Standard to check for quality and this comment applies to all the Standards to which reference is made below.

9.3.2 PVC Soil and Ventilating Pipes and Fittings

The next product to appear on the market, in the early 1960s was PVC soil and ventilating pipes. For waste systems, in contradiction with the European Continent, this country opted for higher softening point materials than PVC such as polypropylene, high density polyethylene, PVC blended with chlorinated PVC, and ABS.

Two problems faced the manufacturer. First, he had to satisfy the responsible authorities that the introduction of a PVC soil system into a domestic building did not create a fire hazard. This was the subject of a very thorough investigation into the behaviour of such systems under conditions of real fire and culminated in an amendment being issued to the Building Regulations in 1973 permitting the use of PVC soil systems inside buildings subject to the conditions specified therein.

The second problem was to establish whether PVC had a softening point high enough to withstand normal domestic effluents. It was found that the hottest discharges came from washing machines. A study of those machines which boiled water showed that they had a maximum capacity of 35 litres (7·7 gal). Further study indicated that boiling water from such machines would discharge at a temperature of 91 °C into the soil stack. Hence the test described in BS 4514 entitled 'elevated temperature cycling requirement' was developed. This test can be briefly described as follows:

The installation consists of a vertical assembly of soil pipe and fittings, complete with support brackets, anchor points and expansion joints. At the top end a moulded boss or branch is fixed to which a short length of waste pipe is attached and through which hot and cold water can be passed alternately. The test consists in passing 35 litres (7·7 gal) of water at a temperature of 91 °C in $1\frac{1}{2}$ minutes, and a rest and drain period of 1 min. This is followed by passing 35 litres of cold water in $1\frac{1}{2}$ minutes and a further drain period of 1 minute. This cycle is repeated 2500 times, representing 50 washing machine cycles per year over 50 years. At the end of the test, the assembly shall be watertight when filled with water to 150 mm (6 in) above the waste inlet and shall accept the passage of a ball 6 mm (0·24 in) in diameter less than the nominal bore of the system.

9.3.3 PVC Underground Drain Pipes and Fittings

A similar test was developed for underground drain pipe and fittings except that the temperature cycling test is combined with external loading. The test is fully described in BS 4660.[9] An assembly of pipe and fittings is buried in a box measuring 0·7 m wide by 1·2 m long by 1·0 m deep (2·3 × 3·9 × 3·3 ft). Gravel is used as the backfill media. A load of 36 kN (8090 lb) for a 110 mm (4·33 in) system and 50 kN (11 240 lb) for a 160 mm (6·3 in) system is applied during the test and the passage of hot and cold water similar to that described above but with the hot water at a temperature of 85 °C carried out, also for 2500 cycles.

The vertical diameter of the system is measured, immediately below the applied load, before starting and on completion of the test and any resultant deformation shall not exceed 5 % of the original nominal diameter.

This test proved extremely useful as a development tool and manufacturers were able to obtain much useful information on the design and thicknesses of their systems. However, the results from this test tend to be inconsistent and can be affected by such variables as the rigidity of the box, the type of gravel used, and the care with which the backfill is placed in the box.

9.3.4 PVC Pipes for Gravity Sewers

With regard to establishing the suitability of uPVC pipes for gravity sewers, the manufacturers decided to take a different approach. Over the years, a number of installations had been carried out and were apparently operating successfully. It was decided to measure any deformations which had occurred.

A total of 14 sites, widely distributed throughout the UK was examined, involving the measurement of some 36 lengths of pipe, varying from 20 to 87 m (65·6 to 285·4 ft). Three of these lengths have been measured twice. Pipes of different thicknesses were included in this study, with diameter: thickness ratios of 44:1, 41:1, 28:1 and 21:1. Pipe diameters ranged from 9 in to 24 in (225 to 600 mm) nominal size and were laid in a variety of soils ranging from sand to clay at depths from 0·6 m to 4·85 m (2 to 16 ft). These pipes were manufactured to BS 3506. The oldest installation measured was 6 years old. A typical recording of the deflection measurements taken is illustrated in Fig. 5.

The conclusions reached as a result of this study were:

1. It is normal to find that substantial parts of the length measured are deformed to only a small extent. It is also normal to find one or two places where the deformation is greater. These deformations are generally due to a change in the resistance exerted by the sidefill or, in other words, places where there is a change in the sidefill resistance due to some anomaly such as insufficient compaction.

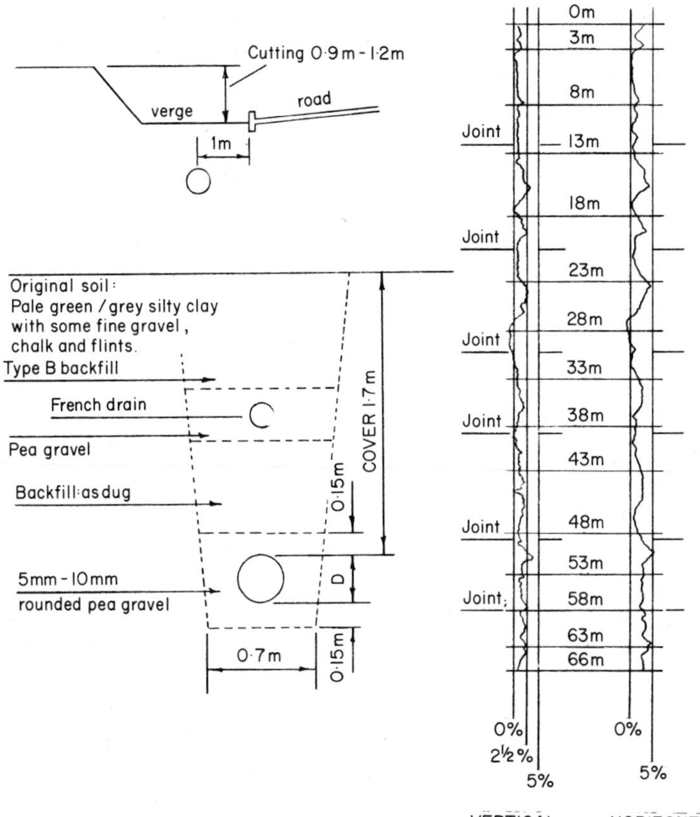

FIG. 5. Typical deflection measurement of sewer pipe.

2. With the exception of three cases, the average deformation is below 5% of the original diameter and in the majority of the lengths measured, the greater part of the lengths was subject to a deformation of the original pipe diameter well below this value.
3. Three lengths have been measured twice at approximately 12 months' interval. The second set of measurements was almost identical to the first.

4. From the results obtained, there would appear to be little difference in deformation due to earth or surcharge loads with pipes laid at varying depths whether under fields or roads.
5. There is some evidence to suggest that misalignment of the pipe line in the vertical plane is associated with high peak deformations. (This may be due to ground settlement and/or incorrect installation.)
6. The average of the peak deformation was in most cases less than twice the general level of deformation.
7. In cases where the enveloping material was reported to have been carefully compacted, the average and maximum deformations were small. In other cases, where the surrounding material was reported as not having been well compacted, the average and maximum deformations were greater.
8. Two sections exceeded a maximum deformation of 10 % with average deformations of 5 % and $7\frac{1}{2}$ %. These sections should be re-examined to see if any reasons for the high peak deformations can be established.
9. Most of the data (22 out of 35 lengths) relate to pipe of Class 41—equivalent to Class B of BS 3505: 1968. In addition, 9 lengths relate to pipe of Class 44—equivalent to Class AA of the preceding edition of this Standard.

These conclusions are extracted from the report prepared by the manufacturers who carried out this work. In the light of this investigation as well as the work carried out by the Transport and Road Research Laboratory,[10] the specification of uPVC gravity sewers, due to be published shortly, defines a range of pipe from 200 mm to 630 mm (7·9 to 24·8 in) with a diameter: thickness ratio of 41:1 for this application.

It is as a result of studies of these kinds that the study of the use of unplasticised PVC pipework has been developed. Experience has since shown that this material is a satisfactory and economic engineering product for a wide range of applications.

9.4 RAW MATERIALS AND MIXING

9.4.1 PVC Types

Of the two methods of polymerisation (suspension and mass) historically mass polymer always had the advantage over suspension polymer in that it had a considerably higher bulk density (610 g/litre). Furthermore, in days before the complete extraction of VCM from polymer as a health requirement, it enjoyed the reputation of being 'easily gelled', particularly on the less sophisticated extrusion equipment of the 1960s. However the most modern suspension polymers now available in the UK have bulk densities of 530–560 g/litre, i.e. approaching that of mass polymer, and the more sophisticated extrusion equipment now available makes the gelation of suspension polymer as easy as that of mass polymer.

A high molecular weight PVC, as defined by its *K* value, is selected for pipemaking to maximise the mechanical properties of the pipe at minimum wall thickness (and therefore cost). The *K* value or specific viscosity of the polymer selected is usually in the range 58–62 (when measured in ethylene dichloride solvent). The melt or extrudate temperature needed to extrude this particular polymer type satisfactorily into pipe is about 185–195 °C. This comparatively low temperature helps to minimise the thermal degradation of the PVC. Additives have, of course, to be added to the PVC in the first or mixing stage of processing.

9.4.2 Additives

The two major types of additives for rigid PVC are thermal stabilisers and lubricants. Thermal stabilisers normally used in the UK for pipe are basic salts of lead (particularly tri-basic lead sulphate and di-basic lead stearate). In America organo-tin stabilisers are used, and in France calcium-zinc complexes are mandatory for water supply pipe. The heavy metals used delay the decomposition which is inevitable under continuous high temperature operation.

The lubricant systems used to assist the PVC powder to absorb heat uniformly in the early zones of the extruder are called internal lubricants and are normally low-temperature melting point waxes. The stearates already present in some stabilisers also assist in lubrication. The second type of lubricants called external lubricants (having a high melting point and which are usually stearates or esters of stearic acid) are used to assist in the rheological process of passing the melted extrudate ('melt') through the die.

Developments over the last ten years, largely induced and ecouraged by the need to minimise the toxic risk of handling lead salts, have led to pipemakers using co-precipitate stabiliser/lubricant systems where the various additives needed in the formula are co-precipitated chemically (see Chapter 4). Again, to reduce the risk to operatives, these materials are now offered in flake or agglomerated non-dusting form, and in nearly all cases are handled automatically, and remotely, in mixing plants.

Other additives used in a pipe-making formulation might include impact modifiers, particularly in the case of 'cold' service, and fillers, usually precipitated calcium carbonate, used to cheapen the cost of the material. The density of fillers and the particle size and shape is such that the savings from using fillers when calculated on a volume–cost basis is lower than on a straight cost basis. However, fillers additionally, even in small quantities (1 %), act as an aid to gelation and an inhibitor of 'plate out' (see Section 9.5.6).

Modifiers are used in certain formulations where there is an elastic requirement in the condition of the melt. These are usually methacrylates, and are used where the melt is formed into corrugations, as in the case of land drainage pipe, or is formed into a shape different from that produced by the die.

The other additive commonly used in formulations for pipe-making is pigment. The one most commonly required is grey, made by mixing carbon black and titanium oxide. Other colours are obtained by using cadmium salts, iron oxide salts, etc., particularly for making browns of various shades.

TABLE 3
A TYPICAL FORMULATION FOR MAKING RIGID PVC PIPE (PARTS BY WEIGHT)

	Using individual additives	*Using co-precipitates*
PVC polymer	100	100
Co-precipitate	—	3·2
Tri-basic lead sulphate	1·2	—
Di-basic lead stearate	0·8	—
Normal lead stearate	0·4	—
Calcium stearate	0·3	—
Wax	0·3	—
Filler	1·0	1·0
Pigment	As required	As required

It can be seen from Table 3 that the majority of the material in the formulation is PVC and that the number of additives is kept as low as possible for reasons of economy and to minimise plate out. The formulation required depends on the quality of the extruder, the die and the extrusion process; this will be discussed later.

9.4.3 High-speed Mixing

In the early years of pipe-making, mixing the raw material ingredients in a cold ribbon blender for about ten minutes was considered adequate. Very soon it was realised that high levels of stabilisation were required to form rigid PVC pipe unless more intimate mixing could be achieved.

High-speed mixers providing a higher degree of shear to the individual components of the mixture and a high velocity to the particles were, therefore, developed. These not only ensured very intimate mixing in a short time, but also the amount of energy imparted to the powder could be used to raise its temperature, thus allowing the lubricants to melt on to the surface of the PVC particles, coating them with the low-temperature lubricants required to assist in the early stages of extrusion. Thus the effectiveness of lubricants increased, and so the amount of lubricant in the formula could be reduced. A second advantage to high-speed mixing was that the stabiliser could also be finely dispersed over the surface of the PVC particles and would adhere to the PVC because of the lubricant film. Finally the pigment, always used in small quantities, would also be very uniformly and finely dispersed throughout the mix. Thus the high-speed mixer became the standard mixer for making PVC blends for extrusion, and these blends were commonly known as dry-blends.

As mentioned previously the energy imparted to a PVC mix by the action of the high-speed mixer raised the temperature of the mix, and it was found that by allowing the mix temperature to rise above 105 °C the bulk density of the powder increased by about 10 %. Since in the 1960s choke feeding of the extruders was common, a higher bulk density dry-blend produced advantages in output rate, since the output rate was a direct function of the ability to fill the back end of the screws. This advantage is marginally less important with more recent equipment, as the dry-blend is metred into extruders in a manner which seldom fills the back end of the screws completely.

After having mixed the raw materials to a temperature of, say, 120 °C, the mix is discharged and transferred to a cooling vessel designed in a form similar to that of a mixer, but having blades with slower rotational speed and the ability to direct the dry-blend to the walls of the cooler, where heat is extracted by means of water jacketing, e.g. a ribbon blender. The mix is normally discharged from the cooler into a dry-blend storage tank or silo at a temperature not greater than 40 °C (see Section 9.4.5). This low temperature is used to ensure that:

1. The powder will not stick together (due to the lubricant film on the particles of PVC not being firm); and

2. As much residual heat as possible is removed from the dry-blend (thus extending its heat life so that it can be stored over long periods without further thermal degradation).

9.4.4 Heat Life

The concepts of 'heat life' and 'heat history' require consideration. PVC, with or without stabiliser and before or after it is converted into pipe, has a certain heat life (above the glass transition temperature), which is a function of the temperature at which it is held and the time at which it is held at that temperature. In the case of a PVC compound the heat life at 180 °C could be about 60 minutes (measured by generation of sufficient hydrochloric acid to change the colour of Congo-red indicator paper). The same compound can be held at a temperature of 150 °C for more than 5 h before degradation becomes measurable. It is obvious, therefore, that everything possible should be done in the processing of PVC to minimise the consumption of this heat life. Heat history can be described as the story of the life of the PVC in temperature–time terms, and gives a measure of how near the material is to its point of degradation.

One of the phenomena of high-speed mixing is that, in addition to the advantages indicated above from high-speed mixing, since all the raw materials are raised to temperatures of above 100 °C, moisture and other volatiles are removed from the powders. This is particularly important where formulations include fillers, because such fillers can be hygroscopic. Residual moisture in dry-blend would show up as small voids in the pipes if it were not removed. The mixing process also removes other volatiles from PVC; low molecular weight polymers, and more particularly residual VCM, are driven off at this stage. It should be pointed out that current legislation requires the removal of virtually all VCM from PVC at the polymerisation stage and raw material suppliers are currently supplying PVC with VCM contents of not greater than 50 ppm.

9.4.5 Electrostatic Charge on Premixes

The high-speed mixing process for PVC produces abrasion between particles and this produces an electrostatic charge on each particle. These charges are mutually repulsive and cause the mixture to have an abnormally low bulk density immediately after mixing. The electrostatic charge on the dry-blend disperses over a period of time (between 12 and 24 hours) and the bulk density then becomes normal for the material. For this reason most pipe-makers keep dry-blend for at least 12 hours after it has been mixed. Suspension polymer and mass polymer both suffer from the same phenomenon, though strangely the charge on mass polymer is opposite to that on suspension polymer, but the effect is the same. However, if a dry-blend of suspension polymer is mixed with a dry-blend of mass polymer shortly after high-speed mixing, the opposite polarities are such that the two mixes will cling together by their electrostatic attraction, and in doing so demonstrate a bulk density in excess of normal. This phenomenon imparts the appearance of 'wetness' and the dry-blend appears similar to wet sand, having the same non-flowing, non-pouring characteristic of this material. Again, when the charge is dissipated, the phenomenon disappears.

9.4.6 Future Usage of Additives

Due to the production of plate out it is desirable to search continuously for improvements in the quality of PVC so as to minimise the amount and number of additives required for processing. A narrow particle size distribution, a uniform particle shape, a very pure PVC and a PVC which has not consumed much of its heat life in the drying process (after polymerisation), all add to the ease of processing with a minimum of lubricant and stabiliser. Regarding the future for additives, the use of toxic stabilisers such as lead is resisted in some countries and non-toxic alternatives, so long as they are as effective as lead, are extremely desirable. The effects of synergy of calcium and zinc salts, or other complexes, when used as stabilisers is being pursued, particularly in France. Tin stabilisers, although not free from toxic risk, have advantages in that they are effective in very small quantities and are stable in the melt. None, however, has the same stabilising quality as basic lead salts.

9.5 EXTRUSION

9.5.1 Twin Screw Extruders

The development of twin screw extruders for the processing of PVC dry-blend has improved over the last 15 years so that such machines are capable of producing at higher output rates whilst adding heat to the material at a rate sufficient to gell it without degradation. Designers of extruders in the early 1960s were limited to low screw speeds because of the problems caused by back pressure. This pressure was absorbed on bearings which were of low thrust capacity due to the small space available between the ends of the intermeshing screws. In addition, the barrel length was limited because the flexure of the screws caused rapid barrel wear. This was at a time when the protection of screws and barrels against wear by hardening was of limited effectiveness. For this reason also the rotational speed of screws was kept below 20 rpm.

Various expedients in the design of thrust bearings—by having them partially in tandem or by using conically-shaped screws such that the backs of the screws were at wider centres than the front—enabled higher loads to be taken on the bearings. The use of a four-screw design in which the barrel was in two separate parts, each with a pair of screws, also enabled higher loads to be taken on the bearings. In this case the material was partially gelled in the upper barrel and was then discharged into the lower barrel for final gelling and metering to the die.

Screw speeds have been increased over the last 15 years, although 9 m/min peripheral speed (about 26 rpm for a 100 mm (4 in) diameter screw) is currently acknowledged to be the limit because of the very high shear rates generated above this speed. Outputs are also limited by the need to produce a quality of gelation in the melt sufficiently homogeneous and uniform to produce a high quality product (see 9.5.4).

9.5.2 Length/Diameter Ratios

The length of the barrel and the screw speed determine the dwell time of any particle in a twin screw extruder, because the extruder can be considered to be a positive displacement pump with the minimum of slip. Since heat has to be added to the powder to raise its temperature sufficiently to melt it, and since pressure has to be applied not only to compensate for the reduction in volume but also to facilitate heating, a longer barrel produces a more tolerable addition of heat. Modern extruders are therefore longer in the barrel than those of 15 years ago (the convention is to refer to the relationship between screw diameter and barrel length). In the early 1960s, barrel lengths of 12 times the diameter of the screws (12D) were common, whereas in the 1970s barrel lengths of 17D are almost the minimum, and some extruders are fitted with 22D barrels. An additional development which has taken place over the last 15 years is the fitting of a vent port on the barrel to which vacuum is applied. This vent is positioned usually in front of Zone 2, measured at a position about five diameters below the hopper inlet for 15D extruders, and positioned about seven diameters in front of the hopper inlet for longer-barrelled extruders (Fig. 6).

9.5.3 Operation of a Twin Screw Vented Extruder

The process of conversion of cold PVC dry-blend into an extrudate of a form satisfactory for shaping into product is as follows. Dry-blend is fed into the extruder usually through some form of dosing unit (normally of a screw

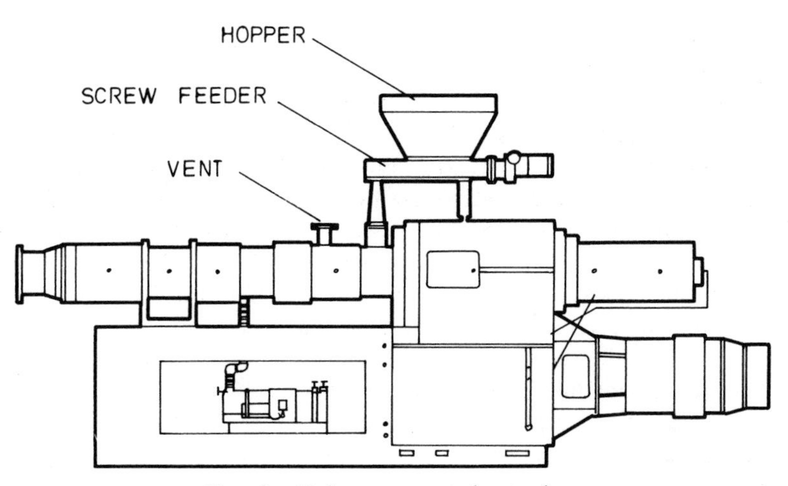

FIG. 6. Twin-screw vented extruder.

feeder type) such that the extruder screw flights at the hopper end are not filled. The extruder screw speed is set to impart energy to the material sufficient to raise its temperature to about 190 °C when the extrudate leaves the extruder. In the first part of its passage down the barrel the powder gains heat from the rubbing motion of particle on particle, but most of its heat at this stage comes from being in contact with the hot barrel.

The barrel of the extruder is heated, usually in four temperature-controlled zones, by electric heaters and Zones 1 and 2 are heated to temperatures between 160 °C and 200 °C. This heat is transferred through the barrel to the powder as it passes through the screws. Leaving Zone 2, the material passes through the vent zone, at which vacuum is applied to draw off residual moisture or volatiles in the dry-blend. In addition the vacuum evacuates air from the barrel, so that none is entrapped in the material as it is carried forward. After the vent zone the material passes through Zones 3 and 4 where compression is applied to the material as well as more heat energy (from the shear effect of the two intermeshing screws).

The compression which takes place in the barrel has to compensate for a change in bulk density of about 0·6 g/cc to that of the specific gravity of PVC (in the compounded form this is about 1·4). Prior to entry into the die, the 'melt' will be at approximately 185 to 195 °C and at a pressure of about 200 atmospheres (20 MN m^{-2}). The pressure in the head of the extruder is induced by a constriction to the flow path of the material. This is done by means of either a small-diameter ring (between the extruder head and the die) called a restrictor ring, or a multi-perforated plate, called a breaker plate. The die itself provides considerable back pressure in addition to this restriction.

9.5.4 Gelation

It can be shown that to produce satisfactory gelation in the material the amount of lubricant added to PVC in the mixing stage has to be balanced so as to control the rate of heat absorption of the PVC in its passage down the barrel. The application of energy to the PVC comes from either electric heaters on the barrel or is supplied in the form of mechanical energy from the motion of the screws. Considering the number of variables in this equation, it is not surprising that extrusion of PVC is still considered to some extent an art rather than a science.

High-quality gelation is required to ensure the development of the mechanical properties of the PVC. This can perhaps be most easily measured by the resistance of a product to impact damage, and in fact impact testing is a standard method for determining the degree of gelation of the material from which the product was made. However, the most important mechanical property of all pipes made from PVC for service under internal pressure is the hoop strength of pipe and it is in this particular test where the quality of gelation becomes most apparent. The test procedure for this (Section 9.2.3) is fairly lengthy. However, there is a chemical method by which the degree of gelation can be shown with some dependability. This is known as the methylene chloride immersion test: methylene chloride attacks under-gelled pipe, showing a white efflorescence in the area where the material is under-gelled.

9.5.5 The Die

In the case of pipe-making the die itself comprises two concentric parts, the outer one called the die ring, the inner one the mandrel. These two parts give the general dimensions to the pipe being made. The die ring and the mandrel are held together by means of a die body and the mandrel itself is suspended in the die body by means of a 'spider' (Fig. 7).

The back pressure generated by the material passing through the channels of the die contributes to the back pressure in the head of the extruder and therefore to the degree of gelation. It is important therefore in designing dies that a proper calculation is made of this effect to ensure that adequate back pressure is generated, but not so much as to inhibit the output rate. The part of the die where it is most simple to vary the back pressure effect is at the parallel part of the aperture. This is known as the 'land' of the die, and the land length is critical. A rule of thumb regarding the land length of dies for rigid PVC pipe manufacture has always been that the length of the land should be between eight and 12 times the width of the aperture. This is known as the land length ratio.

Another simple ratio in die design is that the cross-sectional area at the spider should be about eight times the cross-sectional area of the aperture of the die. This is necessary to ensure the welding together of the melt after it has been split in passing the spider legs. If the weld is not complete a line of

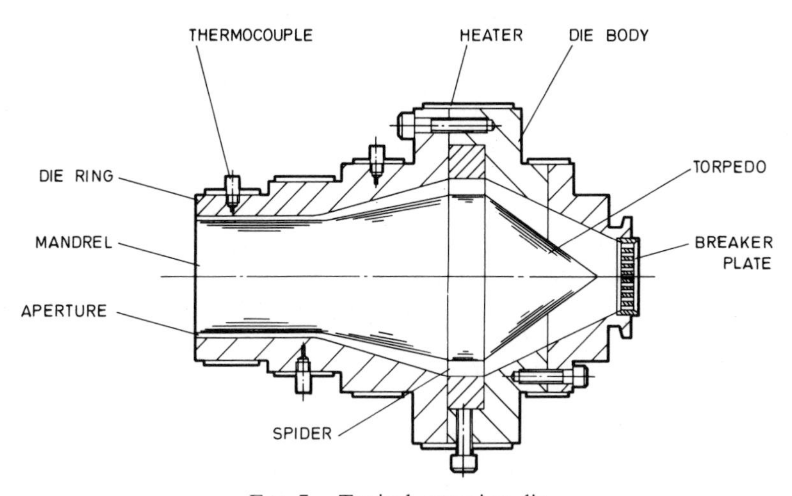

FIG. 7. Typical extrusion die.

weakness is formed which could show up as impact failure, or in the case of water supply pipe, as a split in the pipeline when under pressure. Computer programs are currently in use to determine more exactly the dimensions required in the die to optimise its performance.

However, because of the cost of dies, compromises have to take place in order to minimise the number of dies carried for any range of pipe sizes. For example, in the case of dies for the manufacture of water supply pipe it is quite common to have one die body capable of making pipe sizes between 8 in and 12 in (200–300 mm) diameter and of various wall thicknesses. Obviously not in every case would the spider-to-aperture-area ratio be eight to one. Furthermore, as an added economy, in any one diameter of pressure pipe the same die ring (with the same land length) might be used with different-diameter mandrels to produce different thicknesses of pipe. In this case the land length ratio would not necessarily be optimum. It is in these very proper economies that certain difficulties in producing completely satisfactory pipe can lie.

9.5.6 Plate Out

The phenomenon of plate out has been referred to earlier. This problem usually shows up in the die or in the calibrating box attached to the die. The calibrating box is a tubular extension to the die, cooled with water, on to which the extrudate leaving the die is pressed to initiate the cooling of the pipe and to 'set' the diameter. Plate out, additives exuded to the surface of the extrudate in passing through the die, is deposited usually on the chilled surface of the calibrator, often on the edge of the die tip itself and sometimes within the body of the die. The plate out tends to accumulate and the extrudate is no longer passing over the highly-polished or chrome-plated surface of the die or the interior surface of the calibrator. As the extrudate is still plastic, severe score lines can be produced longitudinally down the pipe. This problem appears in slow-running, large-diameter pressure pipe and the effect is not only that of poor-quality appearance from the scorelines; it can reduce the quality of rubber ring joints. If the grooving is sufficiently severe it can produce a longitudinal line of weakness sufficient to affect the strength of the pipe as a whole.

9.6 POST EXTRUSION OPERATIONS

9.6.1 Cooling

After the pipe has been formed to its leading dimensions, it is cooled. At very low linear speeds it is possible to use a flooded water bath. At higher linear speeds a flooded water bath, with its inefficient heat transfer characteristics, would be extremely long, and therefore spray baths are used.

Recirculating water, cooled usually by air evaporation techniques, is used for cooling the product, and the rate of heat transfer into the water is more influenced by the low coefficient of heat transfer of PVC than by the rate of heat transfer from the exterior wall of the pipe into the water. However, the latter is improved by the use of high-velocity, finely divided sprays, and modern cooling baths are designed to provide this improved cooling. In this way the boundary layer effect of a water film is effectively removed and the best heat transfer available using conventional means is

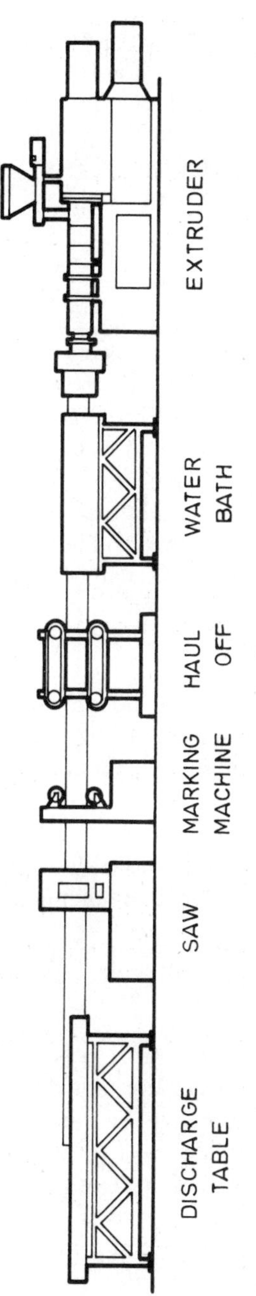

FIG. 8. Typical extrusion line.

achieved. However, for the future in the case of manufacture of thick-walled pipe at high output rates it will be desirable to transfer heat from both the exterior and the interior surfaces of the pipe. Various techniques including extended and cooled mandrels, vertical extrusion (that allows the injection of water into the bore of the pipe as it is extruded downwards) and even evaporation of liquid gases have been considered and tried as aids to extra cooling.

9.6.2 The Haul-off and Other Ancillaries (Fig. 8)

The major ancillary in an extrusion line is the haul-off or take-off, which by providing a constant speed tractive load on the pipe ensures its remaining straight during the cooling process. Furthermore, as the haul-off speed can be adjusted finely, a slightly differential speed between the haul-off and the rate of material extruding from the die tip can 'pull down' the pipe, and in doing so reduce the wall thickness whilst the material is still in a plastic state. By this means control of wall thickness can be achieved. The pipe diameter, set by the calibrator, can to a limited extent be finely adjusted by varying the volume or the temperature of the water cooling the calibrator.

Other ancillaries in an extrusion line usually comprise:

1. A marking machine to mark indelibly the size, class and probably the British Standard number of the pipe being manufactured;
2. A device for determining the length of the pipe, which normally triggers;
3. A saw, or cut-off. In the case of large-diameter pipes (in excess of 6 in (150 mm)) the cut-off is usually of planetary form and often includes a chamfering device for chamfering the ends of pipe, and
4. A discharge table.

9.6.3 End Forming

Virtually all pressure pipe in the larger sizes is coupled by means of rubber ring joints, the housings for which are moulded integrally on one end of each pipe length. The small-bore pipes are occasionally solvent-cemented together, by means of a simple socket formed on the end of each pipe length, into which the spigot of the following pipe length is fitted, with a solvent cement connecting both surfaces chemically.

The ease of use of the rubber ring joint compared with chemically welded sockets recommends it to most users. The equipment for forming the housing for the rubber ring joints has been developed, in the case of pipe of up to 12 in (300 mm) diameter, for use automatically, and for pipe above this size usually semi-manually.

The process of end forming comprises heating the ends of pipe to a uniform temperature of approximately 160 °C. The softened end is pushed over a mandrel, part of which then expands to form a ring groove larger in diameter than the mandrel itself. The mandrel is designed to give a diameter to the interior of the housing slightly in excess of the outside diameter of the pipe. The formed joint is then cooled. By this simple process a joint housing capable of accommodating the spigot end of a pipe and sealing it for the acceptance of internal pressure by means of a rubber ring joint is easily formed.

9.7 THE FUTURE

Future demands for PVC pressure pipe will increase as further penetration of the existing market takes place. Because of the likely disproportionate increase in the cost of plastics raw materials, compared with some of the more traditional materials, it will become more necessary to manufacture pipe economically and to achieve this end it is envisaged that in the early 1980s extruders capable of producing high-quality pipe at the rate of 1000 kg (one ton) per hour will be in use. The output will, in the case of large-diameter pipes, be used to manufacture a single pipe at a time, and to cope with this high output cooling of interior and exterior surfaces will be common. In the case of the manufacture of small-diameter pipe—say up to 6 in (150 mm) diameter—1000 kg per hour of output will be split at the head of the extruder into two streams, each stream making one pipe. Thus, what otherwise would be unacceptably high linear speeds with these output rates, will be accommodated by manufacturing pipes two at a time.

The end forming process will be similar in form to that currently employed, though it is likely that more automatic equipment will be used 'on line', particularly in the case of small-diameter pipe—up to 6 in (150 mm) diameter. In the case of larger-diameter pipe it will be difficult to justify on line end forming equipment because of the amount of space required by the machinery.

Monitoring of pipe wall thickness is possible using ultrasonic equipment and the output from such measuring devices will be used to adjust the concentricity and the thickness of the pipe. This may be achieved by adjustment of the die gap and the haul-off speed.

REFERENCES

1. CP 312—*Code of Practice for Plastics Pipework* (thermoplastics material). Part 1: 1973, General principles and choice of material. Part 2: 1973, Unplasticised PVC pipework for the conveyance of liquids under pressure. Part 3: 1973, Polyethylene pipes for the conveyance of liquids under pressure.
2. PAWSON, J. M. 'PVC Water Mains in Great Britain—The first ten years'. *Water and Water Engineering*, August 1967.
3. NIKLAS, P. H. and EIFFLAENDER, K. 'Results of long time tests on tubes of polyethylene and polyvinylchloride'.
   NIIMAMN, E. and UMMINGER, O. 'Testing of Plastics Tubes'.
   RICHARD, K. and EWALD, R. 'Extrapolation Methods—PVC and Polyethylene Tubes', *Plastics*, April 1959.
4. LLOYD, P. F. V. 'The Testing of Pipes and Fittings', *Plastics*, May 1967.
5. GILL, D. A. 'Long term burst testing for plastics pipes', *British Plastics*, March 1961.
6. GOTHAM, K. V. and HITCH, M. J. 'Design Considerations for fatigue in uPVC pressure pipe', *Pipes and Pipelines*, February 1975.
7. BS 4576 Part 1: 1970. *Specification for Unplasticised PVC Rainwater goods—Half round gutters and circular pipe*, BSI 1970.
8. BS 4514: 1969. *Specification for Unplasticised PVC Soil and Ventilating Pipe, Fittings and Accessories*, BSI 1970.
9. BS 4660: 1973, *Specification for Unplasticised PVC underground drain pipe and fittings*, BSI 1973.
10. OWSTON, C. N. and YOUNG, O. C. 'The use of plastics for non-pressure pipes and underground', TRRL Supplementary report 240, 1976.
11. TROTT, J. J. and GAUNT, J. 'A study of an experimental uPVC pipeline laid beneath a major road during and after construction', *3rd International Plastics Pipe Symposium*, Southampton 1974.

Chapter 10

CALENDERING OF FILM AND SHEETING

R. A. Elden

British Industrial Plastics Ltd, Manningtree, Essex, UK

SUMMARY

This chapter deals with the manufacture of PVC sheeting by the calendering process. The component parts of a modern calendering line are described together with the important constituents of PVC calendering compounds. Methods of accurate thickness and profile control are discussed, as are their effect in the attainment of desired properties in sheeting for particular applications. One can conclude from this chapter that PVC sheeting is capable of production to close specifiable tolerances and finds increasing use as a modern material in its own right.

10.1 INTRODUCTION

The comparative ease with which thermoplastic polymers can be converted into film or sheeting has led to a large increase in their use in the last decade or so. The principal film or sheeting polymers are now polyethylene, polypropylene and polyvinylchloride which have largely replaced the older conventional thermo-polymers nitrocellulose, cellulose acetate, and cellulose acetate butyrate.

The terms 'sheet' and 'film', as applied to thermoplastic polymers, is commonly used to describe all thicknesses of these polymers. Many attempts have been made to define and regularise the use of the two words with little success. For the purposes of this chapter a film is defined as any material less than 75 μm (0·003 in) thickness and a sheet or sheeting as greater than that thickness. Applying this definition of film and sheet an analysis of market statistics shows that polyethylene and polypropylene predominate in the film market whilst PVC dominates the sheet market. The increasing use of polyethylene and polypropylene film produced by the blown bubble extrusion method is seen particularly in the packaging field where these polymers have almost taken over the market traditionally held by cellophane and paper. Whilst these two polymers have found increasing usage over the last few years, perhaps it is the PVC polymer which has undergone the major development in both the manufacture of polymer and the end uses to which compounds made from the polymer have been put. The UK consumption of calendered PVC/sheeting in 1959 was 23 000 tonnes, in 1967 45 000 tonnes and in 1976 67 000 tonnes. The PVC industry was born during the early to middle 1940s with a very narrow range of polymers made principally by the emulsion process. It has now grown to a very large size offering a wide range of polymers having a variety of molecular masses with which are associated a variety of processing characteristics; also a number of monomers can be copolymerised thus adding to the range of properties available. With this development of materials it is now possible to look at a PVC compound as a material in its own right, not merely as a substitute material inferior in properties to the original it replaces. For example, the automobile industry now uses PVC sheeting almost exclusively for the interior trim of motor cars. This position has only been reached by the continuous improvement in the properties of PVC compounds to meet the stringent specification requirements of the motor industry, thus producing a cost effective product. With the development and improvement of PVC polymer and the associated materials necessary for the formulation of PVC compounds, so the markets for PVC sheeting have broadened. Nowadays wide ranges of colours and surface finishes are available including colourless, transparent films in a full range of thickness from 75 μm to 600 μm. Perhaps one of the largest increases in the use of PVC sheeting has been in combination with other materials. This is particularly evident where PVC sheeting is bonded to metallic or wood surfaces to form a decorative and protective barrier to those surfaces.

As a result of these and other developments it is estimated that approximately 20% of all the PVC polymer produced passes through a calender.

10.2 MANUFACTURE OF FILM AND SHEETING

There are two principal methods by which film and sheeting may be produced. These are by extrusion or calendering, and by casting methods, although casting is used only for small quantities. By and large it may be said that most film is produced by an extrusion process and conversely most sheeting, certainly PVC sheeting, is produced by a calendering process, although with such an imprecise definition of film and sheeting this statement is always open to question. For the purposes of this chapter we will consider the manufacture of sheet or sheeting and this inevitably leads us to the calendering process. Figure 1 shows the layout of a typical calendering plant consisting of a number of units. Let us take each of these units in turn.

10.2.1 Premixing

At this stage polymer from bulk storage is fed to a mixer together with plasticiser, also held in bulk storage. The requisite pre-weighed quantities are blended at a temperature of approximately 80 °C. Under these conditions and at this temperature the plasticiser is absorbed into the polymer and a free flowing powder results.

10.2.2 Melt Mixing

The pre-mix is conveyed to the melt mixer which can be either a Banbury or a continuous mixer and thermal stabiliser, lubricant and colour added. The Banbury mixer is a batch mixer capable of producing a batch from 50 kg to 500 kg per charge, depending upon the size of the unit. The continuous mixer produces, as its name suggests, a gelled material continuously, but again according to size can produce at the same hourly output rate as a batch mixer. Both forms of melt mixing equipment are in use, although not in the same plant. The batch mixer lends itself to comparatively short runs where changes of colour and/or formulation are frequent, whilst the continuous mixer gains where long runs without colour and/or formulation changes are needed.

A Banbury mixer is more suitable for gelling plasticised mixes; a continuous mixer will handle plasticised or unplasticised mixes equally well.

10.2.3 Calender Feed

In the past, calenders were strip fed directly from two roll mills but now it is more common to interpose a short barrelled extruder between the mill and calender. This extruder serves two purposes; firstly it provides continuously a feed at a controlled rate which can be easily and accurately varied to match the calender speed, and secondly by the inclusion of suitable gauzes in the extruder head foreign matter can be filtered from the feed.

The provision of mills and extruders at this point in the calender line acts

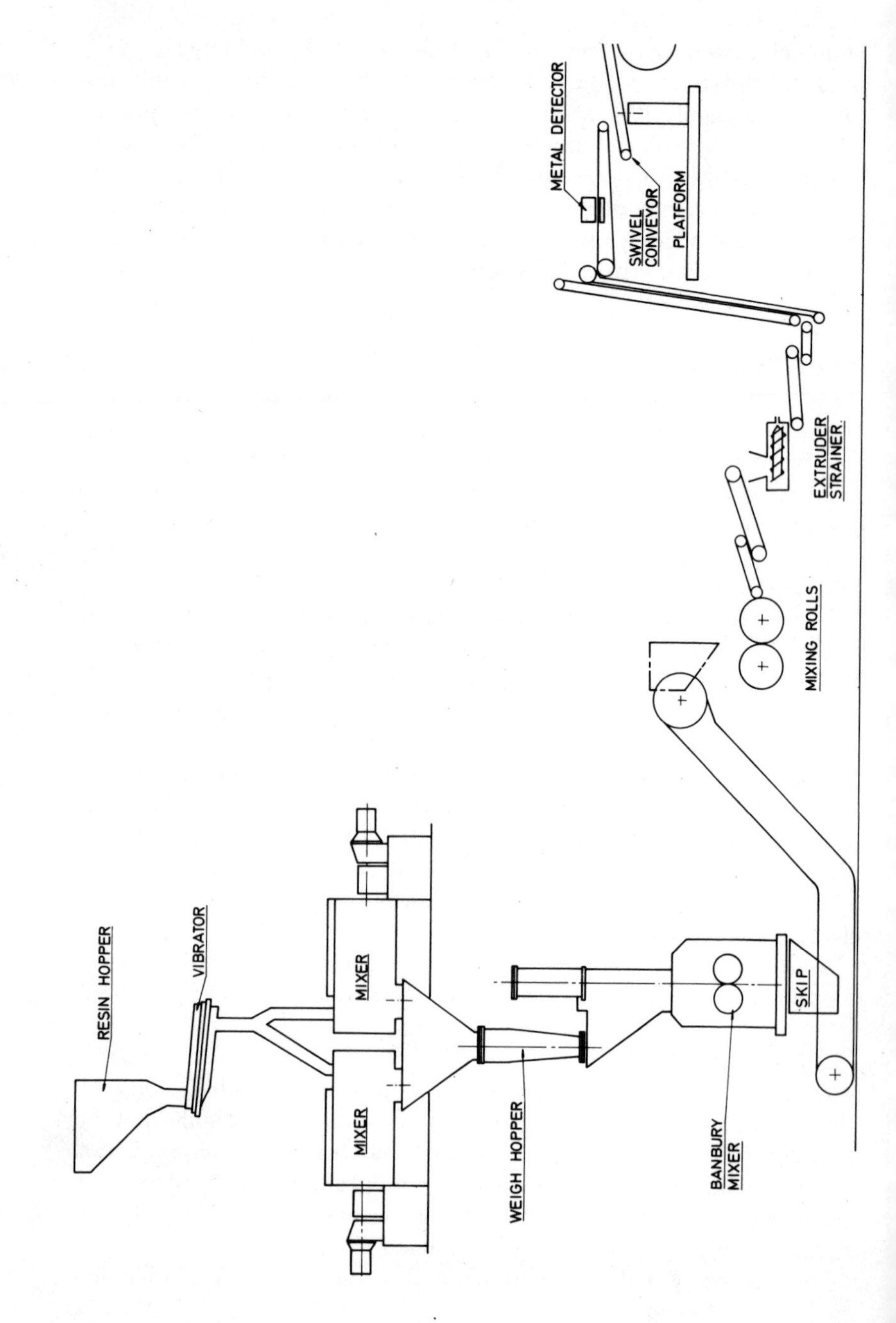
RESIN HOPPER
VIBRATOR
MIXER
MIXER
WEIGH HOPPER
BANBURY MIXER
SKIP
MIXING ROLLS
EXTRUDER STRAINER.
METAL DETECTOR
SWIVEL CONVEYOR
PLATFORM

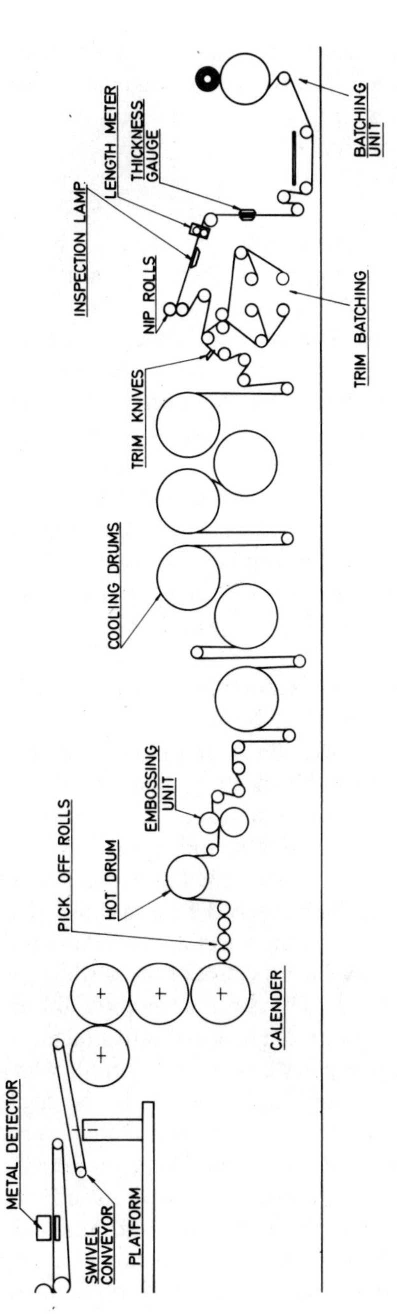

FIG. 1. Layout of typical calendering line.

as a reservoir and thus converts the essentially batch process of the Banbury mixer to the continuous process of a calendering operation. From the extruder the continuous strip is fed via conveyors under metal detectors to the feed nip of the calender.

10.2.4 Calender

New calenders for PVC calendering nowadays are built with four bowls, although five bowl versions for special products such as thinner unplasticised sheeting are in use. Figure 2 shows the common roll configurations employed. The inverted L and Z configurations find most favour in PVC calendering sheet production today. One of the big problems facing the calendered sheet manufacturer is the variety of product demanded which must, for economic reasons, be made on one calender. Ideally a calender should produce one thickness, hardness and width of sheeting but clearly this is not possible. For this reason developments in calender design have been aimed at giving maximum flexibility to any calendering unit. The modern calender has each roll individually driven thus permitting an infinite variation of roll speed ratios. This permits materials with different rheological properties to be made on the same calender.

Calenders with bowl face width up to 2 m are now made, allowing sheeting to be produced which can be subsequently slit to narrower widths if necessary. With wider bowls, bowl deflections can be greater causing thickness variations across the sheeting. Some of this deflection can be accommodated by bowl profiling, but again this method alone is satisfactory for only one hardness, width, etc. Two developments are designed to overcome this problem. The first is roll bending where force is applied to the roll ends to deliberately deflect the bowl. Thus the sum of the deflections on and caused by the sheeting and the applied force can be controlled at a constant level irrespective of the hardness, width, etc., of the sheeting being calendered. The second method of achieving running flexibility is by the application of cross axis adjustment. By this method of control the last calender nip is shaped with a profile which has less crown than might be necessary for sheeting exerting the higher forces in the calender nip; for example harder, thinner material. When roll deflection reaches the point where the sheeting produced is thicker in the centre than at the edges movement is then made to the last two calender bowls so that each roll is moved relative to the other about its horizontal axis. This cross axis adjustment has the effect of increasing thickness at the edges without affecting the centre and affords a method of increasing the product

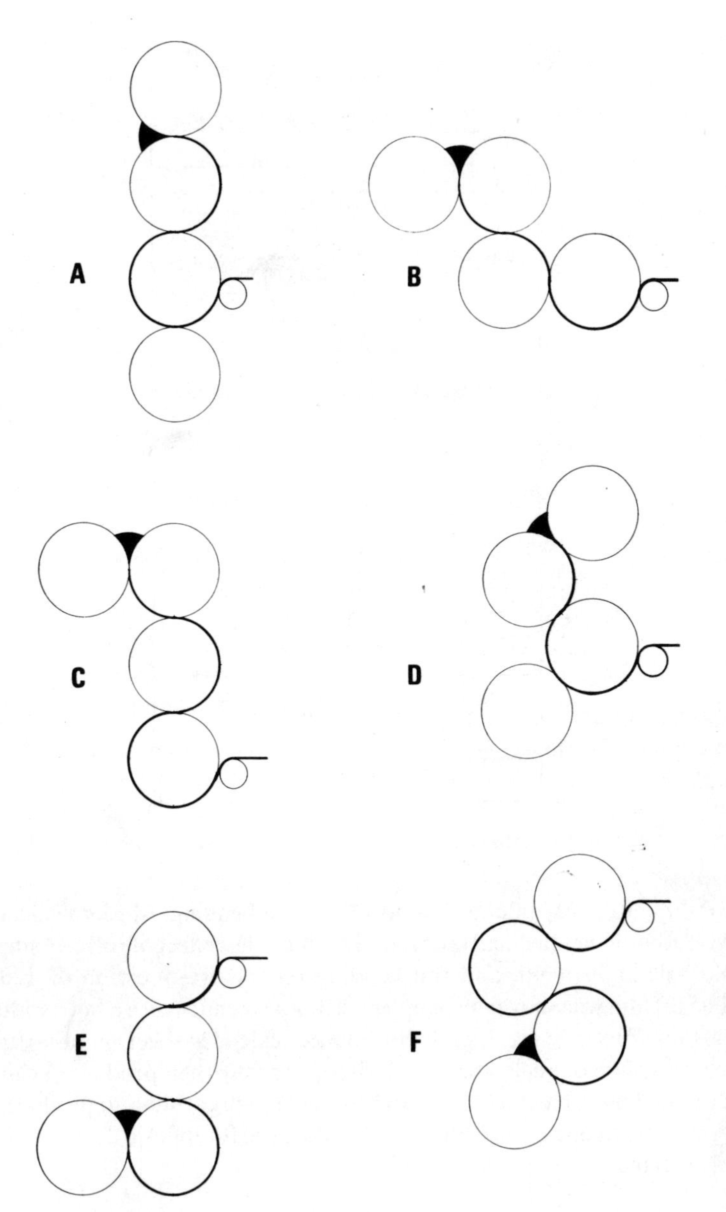

FIG. 2. Typical calender roll configurations; (A) vertical, (B) 'Z' type, (C) inverted 'L', (D) inclined 'Z' (or 'S')-downstack, (E) 'L' type, and (F) inclined 'Z' (or 'S')-up stack.

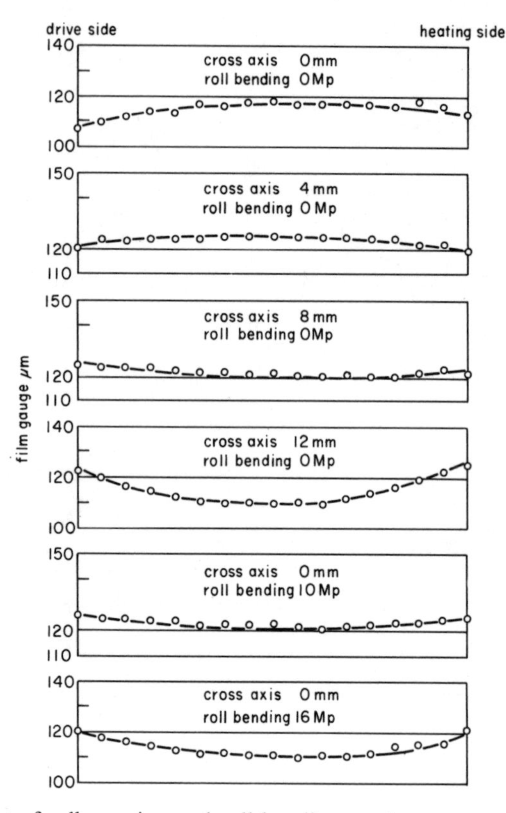

FIG. 3. Effect of roll crossing and roll bending on the cross-section of the film.

flexibility of the calender at less cost than roll bending. Modern calenders have both roll bending and cross axis facilities. The effect of roll crossing, or cross axis adjustment, and roll bending on the cross-section of 120 μm (0·005 in) film was shown by Schuller[1] using a calender with a bowl width of 1800 mm (70·9 in.); see Fig. 3. In practice, calender sheet manufacturers have more than one calender so that they can group their products, again by width, thickness, hardness, etc., and so shape each calender to produce this product within one group thus reducing the need for infinite flexibility from one machine.

10.2.5 Thickness Measurement and Control

As the control of thickness of calendered sheeting has improved, so has the means of continuous measurement of the thickness been developed. At slow running speeds manual measurement with a micrometer was possible but at speeds in excess of 30 m/min this is no longer so and electromechanical methods have been developed. These consist essentially of a low energy radioactive source emitting Beta particles through the sheeting at at least 3 points across the sheeting and causing ionisation in a chamber. The degree of Beta particle transmission varies with the thickness and composition of the sheeting through which it passes. The degree of ionisation and hence the conductivity of the chamber will vary with the thickness for a given composition. These changes in the conductivity can be converted to give a visual read-out of thickness or fed to the calender roll adjusting motors to make continuous adjustments and maintain thickness. With the combination of roll bending, cross axis and thickness measurement of this type it is now possible to calender PVC sheeting to very close thickness tolerances, both along and across the sheeting; $\pm 5\,\mu m$ is not uncommon.

The accurate control of thickness is particularly important to the calender sheeting producer for two reasons; firstly because his customers demand accurate control with no negative tolerance to meet stringent specifications and secondly as a matter of economic viability. Raw materials for calendering are sold by units of weight but sheeting is most usually sold by units of area. Therefore, if a sheeting manufacturer consistently calenders under-thickness he will fail to meet specification, and if he produces over-thickness he will lose money in that he will have to use more raw material to produce a given area. Accurate control of thickness can only be maintained if the roll speeds are constant. For this reason calender manufacturers have devoted a great deal of time and effort to achieving accurate and reproducible roll rotation speeds on both the calender and the take-off.

10.2.6 Cooling and Batching

Having produced sheeting from the calender in the correct thickness it is most important that the sheeting be cooled and wound into rolls without adversely affecting its properties. The sheeting can leave the calender rolls at temperatures in excess of 180 °C and clearly at these temperatures it must be supported and cooled evenly. Cooling trains on calenders are designed to do this with the minimum of longitudinal stretching at temperatures below 100 °C. Should excessive longitudinal stretching take place at temperatures below 100 °C and the sheeting in service reach the temperatures at which stretching occurred then shrinkage takes place in one direction to the detriment of the article manufactured from the sheeting. A typical example of the need for low stretch material is in the automotive industry where PVC sheeting is used for door trimming. It is quite possible during car building or in service for the interior of the car to reach 50–80 °C. If there were more than, say, 2 % stretch in the sheeting at these temperatures, then reversion could take place causing unsightly folds in, for example, the otherwise smooth door trim.

For this reason automotive companies impose very tight specification limits on the stretch permissible in sheeting supplied to them.

Although stretching at temperatures below 100 °C is to be avoided

FIG. 4. Cooling train and contact wind-up of a typical PVC calender. (Photograph reproduced by kind permission of British Industrial Plastics Ltd.)

stretching above this temperature, at, say, 160–180 °C, which is well above the in-use service temperature of PVC, does afford a means of increasing the output flexibility of a calender in that sheeting can be calendered at one thickness and stretched to a thinner one. By calendering at the greater thickness greater output and lower power requirements are achieved. With the correct formulation and conditions a 2·5:1 stretch ratio can be obtained. This means that a calender profiled and roll loaded to calender 250 micron sheeting will produce down to 100 micron sheeting without alteration to the calender profile or power loading.

Having cooled the sheeting it must be wound into rolls for ease of transportation and subsequent use. For flexible (plasticised) PVC, contact batching is very common but this does tend to stretch sheeting at ambient temperatures as it relies on friction between the sheeting and the revolving drum to effect the coiling. The main advantage of this method of batching is its simplicity of operation and low cost. However, because of the need for lower stretch material the use of centre core wind-up units is finding favour with sheeting manufacturers. These consist of a mandrel turned by a constant torque motor and are essential when unplasticised PVC is to be wound up.

Figure 4 shows a typical cooling train and contact wind-up on a modern calender.

10.3 PVC FORMULATION

10.3.1 Polymers

As PVC processing has developed, so have the PVC polymers and the associated materials. In the early days only polymers manufactured by the emulsion process were available. These polymers were comparatively thermally unstable and produced sheeting of poor colour. Now the PVC sheet producer can choose from a wide range of polymers all having excellent thermal stability. Polymers for calendering today are produced by the suspension or mass process and in general terms mass polymers, which are more expensive, offer better thermal stability and therefore less colour development during processing. Because of these properties, mass polymers tend to be used for transparent colourless sheeting whilst suspension polymers of various molecular masses are used for almost all other sheeting applications.

In the most recent past polymer manufacturers have been concerned to develop their polymers to meet the needs of increased productivity from the calender sheet producer. To this end particular attention has been paid to particle size and shape of the polymer and its bulk density. These features are particularly important where bulk handling is employed and polymer must flow freely in the bulk handling equipment. The rate of plasticiser absorption at the pre-mix stage of calendering can be greatly influenced by the physical shape and size of the polymer particles so it is important to the calender operator that faster pre-mixing can be achieved to cope with faster calender speeds without resorting to increased capital expenditure on more or larger pre-mixes.

10.3.2 Stabilisers

In the early days of calendering, PVC compounds were nearly all stabilised against thermal degradation by the addition of lead compounds. However, these compounds had serious disadvantages. Containing lead, they were all either totally inorganic and produced only opaque sheeting or were organic lead salts which reacted with HCl (produced on degradation of PVC) to form lead chloride which at best induced a haze into otherwise clear sheeting. This disadvantage coupled with the toxic nature of lead compounds has led to them being almost totally replaced in calendered sheeting by other metal organic complexes. Of these, complexes of cadmium and barium with or without small quantities of zinc compound are most often used, although compounds or organo-tin are also in favour.

The cadmium–barium compounds are available in liquid form which makes them particularly suitable for bulk handling and automatic dispensing compared with the solid forms of stabiliser. Whilst bulk handling and automatic dispensing has greatly removed the toxic hazards of cadmium and barium compounds to plant operators, it is felt that in the long term these metals are unlikely to be socially acceptable even in the small quantities present in PVC sheeting. For this reason thermal stabilisers based on calcium–zinc complexes in combination with compounds containing epoxy groups are in use and continuing to develop. These stabilisers do not at present have the same ability to retard thermal degradation in PVC as their cadmium–barium counterparts, but even so can be satisfactorily used particularly in the more highly plasticised formulations. Where greater thermal stability is required, for example in rigid formulations, then tin compounds are more suitable.

10.3.3 Plasticisers

Much of the PVC polymer processed is plasticised to give very flexible sheeting. For general purpose sheeting, branch chain phthalates are used since these offer the most economic solution to plasticisation. Attempts to replace these plasticisers with straight chain phthalates derived from C7–C11 alcohols have generally failed since although these latter plasticisers offer improved properties to PVC sheeting by way of better low temperature flexibility and lower volatility, the lower efficiency of them coupled with greater cost does not make them economically attractive for general applications.

Where special properties are demanded of a plasticised PVC sheeting and a premium price can be obtained for such sheeting, a wide range of plasticisers is available to the processor. For example, for low flammability sheeting a range of phosphate plasticisers is available, or where low temperature flexibility is the prime requirement, plasticisers derived from the esters of azeleic acid or sebacic acid can be used and are available. If the principal requirement is for sheeting having a very low volatile loss, then polymeric plasticisers are particularly suitable, although due to their low compatibility with PVC they generally are used in conjunction with monomeric plasticisers such as the phthalates. The plasticiser industry is constantly developing new plasticisers to meet the requirements of the PVC sheeting market, always bearing in mind the cost of the finished product.

The plasticisers mentioned above are all included in the formulation to obtain certain physical properties. There is one class of plasticiser available which is aimed principally at reducing primary formulation costs without detracting too far from the sought-after properties of the sheeting. This is a group known as secondary plasticisers or plasticiser extenders. These are typified by the chlorinated hydrocarbons which have limited compatibility with PVC but when used to replace primary plasticiser at controlled levels can show a saving in formulation costs as they are cheaper than primary plasticiser.

10.3.4 Fillers

In the past inorganic materials such as chalk or china clay were frequently added to PVC sheeting formulations to reduce cost. The addition of these materials invariably impared the physical properties of the sheeting so that today in broad terms the use of fillers is confined to those formulations which require special properties which can only be obtained by the use of fillers. For example, the inclusion of certain fillers in PVC formulations greatly reduces their tendency to burning. Other fillers will improve the impact strength of unplasticised PVC. When using fillers the PVC formulator must always take great care to balance the property sought by inclusion of the filler against the change in general physical properties of the sheeting he is producing.

10.3.5 Lubricants

Hot PVC compound adheres very well to hot metal surfaces. This property allows the calender bowl surface to be impressed upon the surface of the sheeting, but if this adhesion is too high it will become difficult or impossible to remove the sheeting from the bowls. By the addition of lubricants the degree of adhesion to the roll surfaces and hence the surface of the finished sheeting can be controlled.

Lubricants or bowl release agents play a most important part in the manufacture of calendered sheet. They are added purely as processing aids and must have no detrimental effect on the physical properties of the sheet in the short or long term. Typically metallic soaps and long chain waxes are used as lubricants in the manufacture of PVC sheeting.

The choice of lubricant type and quantity varies considerably from calender to calender and from formulation to formulation, and hence the lubrication of PVC sheeting is one area in which the PVC formulator and calender operator must exercise their experience of their own equipment to obtain the desired results.

10.4 POST CALENDERING OPERATIONS

Having calendered PVC sheeting it is easy to enhance its appearance by modifying the surface in subsequent processes. This adds value to the product and can transform an otherwise marginal operation into a more profitable one.

10.4.1 Printing

PVC sheeting is readily roller gravure printed in any number of designs and colours. For most practical purposes, however, up to six colours in one design is all that is required, and two, three or four colours are most common. Most sheeting producers, who also print, rely upon specialist ink manufacturers to supply inks suitable for PVC printing. These must obviously be compounded from pigments which are light-fast and non-fugitive, and as in the case of stabilisers must be free from unacceptable metals.

The inks, usually containing ketone solvents, are deposited onto the sheeting surface from photo engraved rollers and are dried on passing through hot air ovens. After cooling, the printed sheeting is again wound with the same care as from the calender to prevent stretching. Modern multi-colour printing machines are capable of producing printed sheeting at speeds up to 100 m/min (330 ft/min) or more. Therefore, at these production speeds the need for very large calendered rolls is apparent if the economic advantages of fast printing are to be realised. Usually the user of sheeting cannot handle the very large rolls of the producer so the print producer must have rewinding facilities to break down large rolls to smaller, more easily handled lengths.

10.4.2 Embossing

Being thermoplastic, PVC sheeting is easily shaped under the influence of heat and pressure, and is therefore quite readily embossed. This process of passing hot sheeting usually in excess of 140 °C through a nip formed by a mechanically engraved roller and a hard rubber roller can be performed in the calender train or as a separate operation. The order size usually determines where the operation is carried out in that large runs of one colour and embossing can be made on the calender whilst short runs are more suited to a separate operation.

It is sometimes necessary for technical reasons irrespective of order size to emboss as a separate process; for example, where very deep embossings are to be made or temperature of embossings are required which are higher than can be economically achieved at normal calender running speeds.

10.4.3 Stentoring

Much has been said of the need to minimise stretching when roll batching calendered sheeting. However, for certain end uses a controlled stretch at relatively low temperatures in longitudinal and transverse directions is desirable. It is preferred, for example, where PVC sheeting is used for book covering or in loose covering hardboard in automotive trim panels. On heating, the PVC sheeting shrinks in both directions (as opposed to one direction with ordinary calendered sheeting) and a tighter covering results. This process of two way stretching (also referred to as stentoring) is adapted from the textile industry where the process is very common.

Calendered or embossed sheeting is fed through a heated chamber at temperatures below 100 °C and transversely stretched continuously to a controlled amount by expanding grippers. The sheeting is cooled whilst still held in the stretched condition before winding up again into rolls. This process is only applied to plasticised thicker sheeting, that is in excess of 250 microns (0·010 in).

10.5 END USES OF PVC SHEETING

The consumption of PVC sheeting has risen in the last decade and some of this increase has obviously been taken by newer developments in the use of PVC sheeting. This is most evident where sheeting has been used in conjunction with other non-plastic materials. This growth is evident where PVC calendered, embossed and printed sheeting is laminated to wood or metal. In the first case, PVC sheeting laminated to wood, the product is used in furniture production. Furniture today is manufactured from reconstituted wood, that is chipboard, blockboard etc., decorated with a sheet PVC. Such is the skill of the designers and printers that printed woodgrain effects are produced which when laminated are difficult for the layman to distinguish from the natural product, but they have superior properties, not least cost, to the natural product. This is one area which in my view demonstrates that PVC sheeting is a material in its own right, and not an inferior substitute for a natural product. A second growth area for the use of PVC sheeting is the lamination of sheeting to metal. Here the sheeting serves two purposes. First, it provides a decorative finish and secondly, it protects the metal substrate from atmospheric corrosive action. From the decorative aspect, PVC sheeting offers a far greater variety in the form of colour, design and texture than paints which have traditionally coated metals as a decorative finish. These then are two areas in which PVC sheeting finds uses and developing uses, but of course there are many others, for example, in the automotive industry where the physical properties and decorative qualities of PVC sheeting can be used to an advantage.

10.6 CONCLUSION

The production and use of PVC sheeting has increased and developed over the last thirty years and may now be said to have come of age. In the next thirty years will new thermoplastic polymers make their way to the PVC calenders? In this respect, perhaps it is to the olefinic polymers and copolymers that calendered sheet producers should be looking.

REFERENCE

1. SCHULLER, R. Recent developments in the construction and operation of film calenders, part III, *Kunstoffe*, Feb. 1971.

INDEX